BURLEIGH DODDS SCIENCE: INSTANT INSIGHTS

NUMBER 35

# Developing forestry products

I0130694

burleigh dodds
SCIENCE PUBLISHING

Published by Burleigh Dodds Science Publishing Limited
82 High Street, Sawston, Cambridge CB22 3HJ, UK
www.bdspublishing.com

Burleigh Dodds Science Publishing, 1518 Walnut Street, Suite 900, Philadelphia, PA 19102-3406, USA

First published 2021 by Burleigh Dodds Science Publishing Limited
British Library Cataloguing in Publication Data
A catalogue record for this book is available from the British Library

ISBN 978-1-80146-163-4 (Print)
ISBN 978-1-80146-164-1 (ePub)

DOI 10.19103/9781801461641

Typeset by Deanta Global Publishing Services, Dublin, Ireland

# Contents

# Series list

| Title | Series number |
|---|---|
| Sweetpotato | 01 |
| Fusarium in cereals | 02 |
| Vertical farming in horticulture | 03 |
| Nutraceuticals in fruit and vegetables | 04 |
| Climate change, insect pests and invasive species | 05 |
| Metabolic disorders in dairy cattle | 06 |
| Mastitis in dairy cattle | 07 |
| Heat stress in dairy cattle | 08 |
| African swine fever | 09 |
| Pesticide residues in agriculture | 10 |
| Fruit losses and waste | 11 |
| Improving crop nutrient use efficiency | 12 |
| Antibiotics in poultry production | 13 |
| Bone health in poultry | 14 |
| Feather-pecking in poultry | 15 |
| Environmental impact of livestock production | 16 |
| Pre- and probiotics in pig nutrition | 17 |
| Improving piglet welfare | 18 |
| Crop biofortification | 19 |
| Crop rotations | 20 |
| Cover crops | 21 |
| Plant growth-promoting rhizobacteria | 22 |
| Arbuscular mycorrhizal fungi | 23 |
| Nematode pests in agriculture | 24 |
| Drought-resistant crops | 25 |
| Advances in crop disease detection and decision support systems | 26 |
| Mycotoxin detection and control | 27 |
| Mite pests in agriculture | 28 |
| Supporting cereal production in sub-Saharan Africa | 29 |
| Lameness in dairy cattle | 30 |
| Infertility/reproductive disorders in dairy cattle | 31 |
| Antibiotics in pig production | 32 |
| Integrated crop–livestock systems | 33 |
| Genetic modification of crops | 34 |

# Chapter 1

## Developing forestry products: timber

*David Nicholls, USDA Forest Service, USA*

## 1 Introduction

Timber products can include a wide array of socially valued goods originating from diverse tree species, forest types, and ecosystems. Forest product markets are dynamic and constantly changing in response to consumer needs as well as many characteristics of the forest ecosystem, including forest growth characteristics, mortality due to fire and other agents, harvest schedules, and forest policies. Meeting consumer expectations is critical to developing successful markets and wood quality and product quality is central to this. The global forest sector is a dynamic mix of many factors, including wood fiber production, consumption, imports, exports, and prices. At least a dozen distinct product groups in close to 200 nations have strong influence on these macroeconomic factors (Buongiorno et al., 2003).

Over the past few decades, forest product market development has been characterized by several important factors, many of them international. Included are innovation in products and marketing, trade-offs between forest biomass and solid wood, global demands for wood fiber, global macroeconomic forces,

http://dx.doi.org/10.19103/AS.2019.0057.22

Published by Burleigh Dodds Science Publishing Limited, 2020.

the advent of biorefineries, and customization of specialty versus commodity products. This chapter will address each of these, and discuss their impact on global timber markets, as well as future trends.

The twenty-first century could become a turning point for global forest product markets. Some (Hetemaki and Hurmekoski, 2016) have interpreted present markets and events as being in a phase of 'creative destruction'. That is, an era characterized by major declines in some timber-based sectors, with a simultaneous emergence of new products and businesses. The twentieth century witnessed great advancements in the forest sector and timber products modeling, forecasting trends in primary wood products, integrated markets, and increasing levels of international competition and trade (Hetemaki and Hurmekoski, 2016). However, less attention has been given to important emerging issues such as value-added market development, employment, sustainable management, and the diffusion of new products and services into the timber-based marketplace. Many foresee great contributions for the forest sector to meet global sustainability challenges. However, the ever-changing environment for wood products consumption (and therefore timberland investments) will be in response to changing economies, markets, land values and public policies. Some (Toppinen et al., 2018; Jonsson, 2011) have even posited large-scale shifts in timber product capacity and investment growth from Europe and North America to Asia and Latin America.

Others also predict structural changes in the global timber arena. Damaspatra (2009) anticipates that market drivers will include changing demographics and consumer preferences, environmental awareness of 'green products', changes in supply chain management, and integrated trade policies. Still others (Bolkesjo et al., 2006) predict that the development of wood product markets will be closely linked to a parallel development of bioenergy markets. Especially important will be the competitiveness of solid wood manufacturers in relation to changes in commodity energy prices.

Innovation will be needed to respond to dynamic forest conditions. Timber product managers have identified the importance of creating new or improved products, process, and business operating systems as they strengthen market-customer links and build innovative capacity (Hansen et al., 2007). The future development of wood products will also include the 'greening' of products and markets. Environmental marketing, eco-labeling, supply chain optimization, and design for environment are all trends that could all become an important part of wood products and markets in a developing bioeconomy (Hansen et al., 2007). The many aspects of green marketing and green products will likely see increased significance in wood products procurement, supply chain management, production, marketing, and distribution. Innovation will also play a leading role in the future development, adaptation, and success of wood products. Innovation is a widely encompassing concept that can include elements of timber processing, product development, and marketing. Customization, closely related to innovation, is gaining importance in global markets as supply chains and distribution channels become linked between nations, and consumers demand timely delivery of specialized products.

As demand for timber products expands, innovative methods will be needed for management and production of timber products. Currently, products such as cross-laminated timber (CLT) are making inroads, while also reducing greenhouse gas (GHG) emissions when substituted for more carbon-intensive materials such as steel. Not only will more commodity lumber products be needed to meet the basic needs of a growing population, but new specialty uses and applications will emerge.

As more plantation wood is grown, unique applications will necessitate further evaluations of harvest schedules, wood properties, and financial evaluations. Although each region, species, and ecosystem could offer unique opportunities, some generalizations can guide future plantation management decisions. For example, attractive financial returns have been estimated for exotic species plantations in China, South Africa, New Zealand, Indonesia, and the United States, ranging from 7% to 12% annually (Cubbage et al., 2010). The same authors found that rates of return greater than 20% were possible, mostly under eucalyptus plantations. However, certain nations have been identified as having greater risk factors for forest-based businesses and international trade, which could influence financial returns associated with timber products as well as the ability to meet global wood demands. Some of these risk factors include political factors, commercial factors, wars, and the ease of doing business with governments (Cubbage et al., 2010). Developing appropriate and successful timber-based products will be strongly influenced by numerous related issues. These include climate change mitigation, use of biomass for energy, coproduction of wood and non-timber products, and other forest values such as recreation, water use, and wildfire risk mitigation.

## 2 The role of solid wood in the emerging bioeconomy

### 2.1 Solid wood products versus wood energy

Forest harvests often include merchantable stems, from which sawlogs, small-diameter stems, limbs, and foliage are derived. Thus, economic timber utilization is strongly linked to all merchantable components of the tree stem. Wood products market development can occur in conjunction with biomass energy development and biofuels if both types of materials are effectively utilized. However, producers and consumers must find the right balance of solid wood products and biomass to keep pace with technological change while accurately reflecting consumer needs under sustainable forest management. Integrating wood fiber supply chains with resources and products will be central to this objective. However, the future course of energy prices, subsidies, and incentives, are all unknown factors which could influence the balance between products, creating greater demands for woody biomass when alternative energy prices are high, while lessening the demand for solid wood products. Thus, broad-based interactions between timber, forest products, and bioenergy products will be present, and must be accounted for in this segment of the bioeconomy.

## 2.2 The bioeconomy

The term 'bioeconomy' has no precise definition, yet may be best understood by its societal benefits. Wolfshlener et al. (2016) assert that European forests play an integral role in the bioeconomy by providing materials (wood and non-wood), various bioenergy and biofuels products, and a 'wealth of other regulating and cultural ecosystem services'. Bugge et al. (2016) argue that the meaning of the bioeconomy is 'still in flux', but that a central aim is one of bridging the gap between the increased use of bio-technology and the increased use of biological materials. The emerging bioeconomy offers the potential for major $CO_2$ reductions, increased employment (especially within rural economies), and wildfire risk reductions among other benefits. However, the integrated biorefinery is still in its early stages of development. Further, economies of scale could require large volumes of woody biomass and face challenging transportation logistics. The high costs associated with biomass logistics have been identified as a key factor limiting biorefinery investment (Kircher, 2014). Integrated forest biorefineries could encompass several different technology choices, including chemical, thermochemical, or biochemical. Diverse partners would also be needed outside of the sphere of wood product manufacturing firms, among them forest landonwers, chemical producers, and/or technology providers (Janssen et al., 2008).

   Although forest biorefineries can include use of cellulosic woody biomass, a number of other feedstocks and configurations are possible, especially when also considering agricultural residues as a feedstock. Included are first-generation biorefineries (sugars and starches), second generation (non-food biomass including corn stover), third generation (algae), and fourth generation (used vegetable oil) (Batsy et al., 2013). Biorefineries can also produce a spectrum of bio-based products (food, feed, chemicals, materials) and bioenergy (biofuels, power, and/or heat) via the sustainable processing of biomass (IEA Task 42). Thus, the bioeconomy could encompass wide-ranging feedstocks, technologies, material pathways, scales of operation,

Published by Burleigh Dodds Science Publishing Limited, 2020.

and stakeholders. Optimizing the use of diverse cellulosic feedstocks could prove challenging. For example, Parker et al. (2010) considered biorefinery site optimization across the western United States, using mixed integer-linear optimization models, and including woody and agricultural residues. They found that biorefineries could collectively provide close to 15% of regional liquid transportation fuels.

As the bioeconomy develops, timber utilization will be closely linked to use of other material feedstocks. For example, plantation forests could be managed to produce short rotation-intensive stands for bioenergy products, or could be managed solely for including merchantable stems for sawtimber. The economic trade-offs between production of commodity lumber, commodity biofuels, specialty wood products, and specialty biofuels will be dynamic, and closely related to commodity prices. In addition to solid wood and biomass feedstocks discussed earlier, the developing bioeconomy will encompass many feedstocks from both forest and agricultural regions. Enhancement and commercialization of tree products has been identified as a new frontier in agroforestry, as articulated in the Millennium Development goals (Garrity, 2004). Complex, interrelated supply chains will be needed to ensure timely delivery of feedstocks to production facilities. Supply chains that are likely to become more fully developed include feedstocks from not only forest lands, but agricultural lands. Barriers to future bioeconomy development include the high costs of biomass harvest and transportation, as well as the ability to scale smaller biorefineries with the large scale now used in petrochemical refineries (Kircher, 2014). A thriving bioeconomy will involve trade-offs between raw material utilization, processes, product types, and markets, while also enhancing sustainability.

## 3 Sustainable timber products: wood product use and carbon sequestration

### 3.1 Forest carbon

The potential for global timber to provide environmental and societal benefits is extensive, and one area that is gaining importance is the carbon benefits provided by forest ecosystems. Forests carbon benefits can accrue from several sources, including on-site carbon sequestration, solid wood utilization, and bioenergy. However, determining carbon flux estimates and net changes in carbon stocks in actual forest ecosystems is complex. Included in this web of variables are changes in soil carbon, levels of merchantable timber harvest, small-diameter stem removal, use of forest harvesting residues, forest regrowth, production of durable wood products, bioenergy use, and substitution of wood for more energy-intensive materials (such as concrete and steel). Forest carbon reserves are substantial. Managed forests are predicted to store up to 184 Terragrams (Tg) of carbon per year (over the period 2000-2050) when also considering the carbon sequestered in durable wood products (Sohngen and Sedjo, 2000). In similar work, Pingoud et al. (2010) evaluated equilibrium states of managed forests and wood product pools to reflect long-term sustainable strategies. They found that climate benefits appear to be highest when sawlog supplies are directed toward long-lived products that substitute for more energy-intensive materials. Others (Jansenns et al., 2003) have estimated that forest growth in Europe sequesters between 7% and 12% of anthropogenic carbon emissions.

Sohngen and Sedjo (2000) presented carbon flux estimates due to increased demand for timber products from industrial forests. An additional 184 Terragrams (Tg) per year of carbon are projected to be stored in forests and wood products over the first half of the twenty-first century. They suggest that, although harvests in natural boreal and tropical forest regions will cause carbon releases, new plantation subtropical regions could more than offset these losses. In pioneering research, Winjum et al. (1998) proposed two approaches for measuring GHGs at the national level: an 'atmospheric flow' method (using the atmosphere as a frame of reference) and a 'stock-change' (using forest and wood product stocks as a reference). In doing so, they estimated total global carbon stocks related to timber products utilization to be 980 Tg carbon in 1990. The distinction between developed nations and developing nations was cited as an important variable in the global carbon budget.

When considering the carbon benefits of forests, substantially different life cycle impacts can be attributed to single stands versus larger forested areas, and the specific carbon pools involved. Raymer et al. (2009) found that there are trade-offs between maximizing revenues and maximizing carbon benefits in regional forest tracts in Norway. Using dynamic optimization, they found that management strategies that maximize carbon benefit can decrease the net present value (NPV) of regional timber revenues by 21%. However, when substitution effects were also considered, the NPV of carbon benefit increased by a factor of about 1.5.

## 3.2 Carbon fluxes in forest stands

A key aspect of accurately predicting global net carbon sequestration is that different geographic regions vary in their carbon flows, with the potential for some regions acting as forest carbon sinks with other regions carbon sources. The global levels of forest carbon sequestration versus emissions for boreal, temperate, and tropical forest growth must be considered in determining global carbon fluxes. Further, the interactions between actively managed forests versus passively managed natural forests versus plantation forests, the impacts of land-use changes, and the potential for forest land area to increase or decrease over the long term can influence carbon flows and net changes in forest ecosystem carbon stocks.

The inherent impacts of climate change and its influence on forest productivity are already underway. For example, warming temperatures and longer growing seasons could have a positive influence on forest growth, at least in the short term; however, increased wildfire risks could offset these gains. The trade-offs between forest carbon (whether in standing trees, forest biomass, and soil) versus in durable wood products will be guided by social values which could be motivated by wildfire risk mitigation. This will help determine specific carbon flows that can enhance sustainability as well as economic metrics of timber-derived products.

## 3.3 Carbon in durable wood products

Changes in net carbon stocks and emissions are important components of country-level assessments of GHGs. Using methods based on the UN convention on climate change and the FAO global forest products database, Winjum et al. (1998) found that developed nations versus developing nations contribute almost equally to wood products carbon stocks. Further, durable timber products have the potential to reduce life cycle carbon emissions, particularly long-lived products (Sathre and Gustavsson, 2006). Forest carbon management and wood products carbon management are closely linked in their roles in climate change mitigation (Hennigar et al., 2008). By optimizing carbon storage in these pools, scenarios for maximizing storage benefits over a 300-year planning horizon are possible. These findings reinforce the importance of including products in forest-sector carbon budgets.

Life cycle assessment (LCA) is used to account for material and energy flows at all stages of a product's life, including timber products. In the case of carbon within forest ecosystems (including stems, branches, and roots) versus carbon in manufactured products (such as lumber and wood composites), LCA can be used effectively at all stages including growth of standing timber, harvesting, transportation, processing, end-use, and disposal. The net carbon sequestration benefits of long-lived wood products (for example in residences and other durable structures) must be reduced by the carbon emissions due to harvesting, transportation, processing, and other life cycle events. Whether timber is processed into lumber products, wood composites, or energy products can greatly influence its life cycle emissions and carbon footprint.

For example, Puettmann et al. (2010) documented cradle-to-gate life cycle inventories for several timber-based products produced in the United States. Life cycle energy consumption for producing softwood lumber was only about 50% of that for hardwood lumber products. Hardwood lumber and flooring manufacture had relatively high energy requirements due to heat generation for drying, which was in turn influenced by higher initial wood moisture, denser wood, and longer, slower kiln-drying schedules versus softwood lumber (Puettmann et al., 2010). Further, the use of biomass as an energy source for manufacturing could greatly reduce life cycle emissions if it is substituted for fossil fuels. Timber product use is important not only from the standpoint of engineering design considerations (i.e. material properties) but also whether a less carbon-intensive material could be used while still meeting other design specifications (Bowyer et al., 2005b).

Carbon emissions during manufacture of building materials can also be a significant. For example, softwood lumber carbon emissions during manufacture can average 33 kg per metric ton, much less than emissions from steel production (220 kg per metric ton), or concrete production (265 kg per metric ton) (Crampton, 2017). Additional benefits can be derived over the life of a timber structure (Kaziolas et al., 2015) especially if there is flexibility in the heating source and design inside temperature of the structure. This is due to the insulating abilities of wood being greater than either steel or concrete, resulting in reduced heating needs.

However, when producing different timber products containing numerous components, there are many integrated carbon pools involved, and thus complex methodologies and results. Bowyer et al. (2005b) compared wood building materials versus steel and also versus concrete, also finding life cycle carbon flows and overall emissions being lower for wood.

Advanced modeling and optimization techniques can be used to accurately assess more complex durable wood products. Kaziolas et al. (2015) used simulated annealing and genetic algorithms to perform an LCA optimization of all-timber buildings. For timber structures, LCAs must account not only for energy consumed in harvest, production, and use of building materials, but also the energy consumption during the lifetime of the structure. Wood is noted for its high insulating value versus other building materials such as steel. Thus, a complete life cycle inventory must account for the potential for reduced heating requirements of an all-wood structure during its lifetime. In some cases, timber structures may result in lower lifetime heating requirements, creating additional carbon benefits (especially when fossil fuels are used).

In other cases, product substitution can lead to more favorable carbon flows. Life cycle comparisons made between wood and other building materials (notably concrete and/or steel) have quantified the full 'cradle-to-grave' life cycle benefits (Bowyer et al., 2005b). Eriksson et al. (2012) developed integrated models for forests and solid wood products in the EU, concluding that increased levels of wood construction could be an effective tool for carbon sequestration, while having minimal impacts on forest ecosystem carbon.

## 3.4 Cascaded wood products

The 'cascade' principle suggests that wood material is prioritized based on the added value that can be generated, so that preferential use would be products with long life span. Products with short life spans such as bioenergy would be derived from waste wood, wood residues, or recycled products. The use of wood fiber for energy (after recycling opportunities have been exhausted) is considered as the least valuable option (Ciccarese et al., 2014).

Future uses of timber products in a global economy will require efficient use of wood fiber to meet demands in diverse markets, locations, and demographic groups. In a simple example, solid wood can be used for lumber, followed by use as reconstituted products, followed by use for energy. Cascaded products are used in various 'incarnations', as different products at different life cycle stages. Perhaps the greatest potential for timber products in mitigating climate change will be in cascaded product chains. Here wood fiber is used more than once, in different forms, with each successive product having distinct uses (and therefore carbon benefits). The potential for greater emission reductions is apparent when solid wood is utilized initially. For example, lumber products in residential construction can be salvaged at the end of their useful life, chipped, and then burned for energy. In other cases, timber products can be cascaded (i.e. re-purposed) three or more times. For example, solid wood products could be reconstituted at the end of their useful life and used to produce wood composite boards that would eventually be burned for energy.

By cascading wood products, life cycle global warming potential can often be reduced. This is especially apparent when wood products are substituted for products requiring greater fossil energy intensity – for example, steel or concrete (Skullestad et al., 2016). Hoglmeier et al. (2015) developed algebraic LCA models to account for material flows when producing various cascaded wood product series. They found that, due to the efficiencies of product cascading, raw material savings of up to 14% were possible. Thus, product cascading can have significant effects on management of timber supplies.

Bais-Moleman et al. (2018) evaluated the use of biomass as a cascaded material under several LCA scenarios in likely wood production chains. By quantifying overall wood flow within the EU forest and bioeconomy sectors, they found substantial benefits from cascaded use. The overall efficiency of wood use could be increased within a range of 23-31%, and cradle-to-gate GHGs could be reduced within a range of 42-52%.

All-timber multistory buildings are constructed of lumber, CLT, and other solid wood components. Thus, they are comprised of 'minimally processed' wood, and they occupy the first position in a cascaded product series. Using attributional LCA methods, Skullestad et al. (2016) evaluated timber multistory structures ranging from 3 to 21 stories. They found climate change benefits ranging from 34% to 84% (versus reinforced concrete structures having similar design criteria), with taller wood structures realizing greater environmental benefits.

Hildebrandt et al. (2017) estimate that increased use of engineered wood products and timber structures in Europe could result in net carbon storage of about 46 million tonnes $CO_2$ equiv. per year by 2030. Much of this benefit comes from reduced use of concrete and/or steel. However, new policy instruments would need to favor the voluntary use of innovative wood construction materials, to reach this goal. A major part of this will likely be timber use in multistory construction. Gustavsson et al. (2010) quantified LCA energy use and carbon emissions for an eight-story wood apartment structure, finding that negative life cycle $CO_2$ emissions can be achieved when considering the carbon benefits of both the wood-based construction materials and the biomass-based energy supply system.

Sikkema et al. (2017) point out that efficient cascaded wood product schemes would reduce (or even prevent) the first intake of high-quality wood fiber for energy or paper products. There are currently economic incentives to do so, since in common practice large trees above a certain diameter are first used for sawnwood or veneer; with remaining wood residues used for pulp and paper products or for energy (Sikkema et al., 2017). Thus, optimum cascading would involve long-lived solid wood products being produced from high-quality harvested logs (with later uses including reconstituted wood products, and finally energy products).

Pingoud et al. (2010) evaluated the dual climate benefits of long-lived timber products in Finland, including the combined carbon stock in forests and wood products plus the fossil carbon emissions displaced by harvested wood products. They found that the climate benefits appear to be highest when sawlogs are directed to production of long-lived materials substituting for fossil-emission and energy-intensive materials and recycled after their useful life to bioenergy. By cascading wood products, global warming potential can be reduced, in some cases by up to 14%. This is especially apparent when wood products are substituted for products requiring greater fossil energy intensity (e.g. steel). However, there is a strong need for more efficient and cost-effective recycling in the later cascading stages (Sikkema et al., 2017).

Published by Burleigh Dodds Science Publishing Limited, 2020.

# 4 Innovativeness in new timber-based products and processes

## 4.1 What is innovation?

Innovation has been studied by many researchers across many business types and business models, yet remains difficult to define in all contexts. Innovation can occur within products, processes, and business systems, and management culture (Hovgaard and Hansen, 2004). In central Europe, Kubeczko et al. (2006) found that when considering product and service innovations to support the performance of forest holdings, innovation can create new income and employment and therefore contribute to rural development. Innovation is also often closely linked to competitiveness. Within the forest products sector, industry managers have consistently indicated five elements of innovation (Hansen et al., 2007):

- being 'new' (or current)
- creating the 'right' culture
- managing the links between markets and customers
- being a leader, and
- focusing on the future.

Although there is a general acceptance that innovation can positively influence competitiveness in the forest products industry, there is little research or hard data verifying this (Hansen, 2010). Successful market development of timber products will require innovation to identify new products that are of high quality and meet important social needs. Developing new markets and products will be integral to the advancement and future success of global timber products companies. For example, Hansen et al. (2014) summarized innovation research in the North American forest sector, identifying seven key themes. Many of these relate to new product development, corporate culture with regard to innovation, and objective means of measuring innovation. An overarching goal of this work was to relate innovation to firm performance.

The lumber industry has been perceived as generally lacking innovation, and for North American producers, a 'commodity mentality' has often been motivated by low-cost strategies for wood fiber recovery rather than innovation. Crespell et al. (2006) found that sawmills having high levels of innovation were also effective in product marketing, and that highly innovative mills tended to have greater investment in new product development processes. Wagner and Hansen (2005) found that firm size does impact the innovation type pursued by wood products companies. Here, large companies tended to outperform smaller companies in process innovation. However, small companies were more evenly matched with larger firms when considering all three innovation types (process, product, and business systems). These findings are particularly important within the softwood lumber industry, given that it is a primary producer and is often closely related to production of CLT and other timber-derived products. Further, it illustrates that product- and process-related innovation can occur not only at the management level, but also at the production level within sawmills.

Innovation and innovative capacity can also be used to better adapt to economic downturns (Hansen, 2014). For example, the financial crisis of 2008-09 resulted in decreased demand for wood fiber, fluctuating prices and exchange rates, and increased competition for many wood product firms. However, innovative companies were able to remain resilient, creating competitive opportunities (Panwar et al., 2012). Thus, innovation enabled companies to recover from recessions more adeptly than the forest sector as a whole.

## 4.2 Wood products clusters

Business cluster can be defined as 'geographic concentrations of interconnected companies … firms in related industries … or in a particular field that compete but also cooperate' (Porter, 2000). Wood product business clusters typically are characterized by geographic proximity, similar product types, and often a shared supply chain network. Effective timber-based business clusters can spur innovation and the diffusion of new ideas and business practices. Business clusters often enjoy synergistic advantages that can help create shared value (Kramer and Porter, 2011). Among the potential advantages are lowered waste and emissions costs, more efficient materials procurement and distribution, and improved financial performance for individual firms and for the cluster as a whole.

Innovative bioeconomy business clusters include a wide variety of products, processes, and conversion technologies (Scarlat et al., 2015). The degree of processing and conversion can range from relatively simple for timber-based solid wood products, to advanced technologies for products such as liquid fuels or nano-scale products. Innovation is also needed for efficient functioning of clusters and to ensure year-round feedstock supplies, efficient conversion processes, and efficient flow of energy and materials between cluster businesses. Strategic decisions regarding the sustainability, location, and operation of a bioproducts cluster must consider all upstream and downstream processes and energy and material flows (Elia and Floudas, 2014). A major challenge in bioproduct cluster development, including for timber products, can be the combination of complex conversion processes with uncertainty in biomass source and supply (Hong et al., 2016). For example, certain materials could represent a waste product to some industries while being a valuable raw material to others. Ideally, wood product clusters would include some start-up businesses in addition to mature firms, as well as a range of businesses sizes, providing additional diversification.

Clusters characterized by creative networks and that promote research, development, and innovation can provide wide-ranging economic and environmental benefits (Margareta, 2014). The concept of clusters has become more and more popular in recent years; nevertheless, the idea of cluster research in the wood processing industry is relatively new.

## 4.3 Innovation case study: cross-laminated timber (CLT)

Several product classes stand out as being innovative new uses of timber. Cross-laminated timber (CLT) has been developed in Europe over the past 25 years (and more recently in North America over the past decade), and is finding use in several products, most notably multistory structures constructed primarily or entirely of wood. Over the past several decades, mass timber structures are gaining importance as an innovative use of wood. Even though they employ relatively simple and well-established construction techniques, innovative uses of small-diameter and diseased timber are helping to find markets for otherwise unusable wood fiber, while reducing wildfire risk. Further, CLT-based structures have the potential for long-term carbon sequestration, among other life cycle benefits.

In CLT products, several perpendicular layers of sawn wood are manufactured into alternating layers, with glued-up construction creating a finished product not unlike a thick plywood panel. CLT dimensions can approach 16 meters long and 0.5 meters thick; however, many configurations are possible when manufacturing individual members (Van De Kuilen et al., 2011). CLT has been used successfully in high-rise construction of up to 18 stories (Fast et al., 2016), and comprises close to 23 structures internationally of at least seven stories (Bowyer et al., 2016). As early as the 1990s, multistory timber construction was under development in Finland as a means of more sustainable construction versus traditional methods (Riala and Ilola, 2014). Much of this innovation has been spurred by opportunities to advance timber use while reducing carbon footprints of buildings. For example, Skullestad et al. (2016) applied attributional LCA to model the climate change benefits of all-timber structures. They found significant benefits, especially for taller (12- to 21-story) timber structures. Waugh et al. (2010) estimated that 169 tonnes of carbon would be stored within an eight-story all-timber spruce structure during its lifetime.

Many innovations are being incorporated into the construction of CLT structures. Developments in the design of mechanical joints, integrated steel tension cables, and glued connections are allowing for greater shapes and configurations of CLT high-rise timber structures (Van De Kuilen et al., 2011). Innovative construction methods are also being used to limit creep movement and moisture-related change in all-timber structures (Waugh et al., 2010), important factors when comparing timber to alternative building materials over long time frames. Further innovations allow the use of timber combined with concrete, where the timber component would be close to 80% of building materials. In certain cases, 'wood-concrete skyscrapers' may be feasible, consisting of concrete cores surrounded by structural timber elements. In this scenario, very tall structures of up to 150 meters (approximately 48 stories) may be possible, of which 80% or more may be timber (Van De Kuilen et al., 2011). An important advantage of timber-based construction in high rises is the time and cost savings of using primarily wood versus traditional materials such as concrete. In other cases, hybrid structures are gaining prominence. Wang et al. (2014) found that, in UK markets, green building trends increasingly include the use of hybrid structures (such as wood and steel) or composites (such as wood and plastic). Although Scandinavian nations are leading the way in this arena, more research is needed to quantify the environmental benefits of increased use of wood as a building material (Toppinen et al., 2018). Hurmekoski et al. (2015) found that diffusion of multistory wood construction will be facilitated by the construction value chain becoming less risk-averse toward the adoption of new practices.

### 4.4 Innovation summary

Innovation is a concept that offers no precise definition; however, a central characteristic of innovation is the need to be current and adaptable to change. Yet it is central to most successful companies, especially with regard to timber-based products. However, many areas within the forest sector tend to be conservative in their business practices, failing to invest sufficiently in innovation and new product development (Hansen et al., 2014). Due to the inherent variation in raw material, including species, size class, wood quality, and moisture content, the production of timber products and by-products requires innovation at every step of manufacture. Further, as consumer markets become more specialized, product customization will become more important. Therefore, there is a clear need to find balance between timber-based commodity products and specialty products. Innovation in products, processes, marketing, business systems, and other areas is likely to play a key role in future development of global timber markets (Hansen et al., 2007).

Business clusters often enjoy synergies that can help create shared value, including lower waste and emissions costs and more efficient materials utilization. Innovative bioeconomy business clusters include a wide variety of products, processes, and conversion technologies (Scarlat et al., 2015).

Published by Burleigh Dodds Science Publishing Limited, 2020.

Wood product clusters characterized by creative networks can foster greater levels of innovation, providing wide-ranging economic benefits. However, as the number of timber and bioeconomy products expands, reliable biomass supply chains will be a critical need considering the complex conversion processes and material flows involved.

Innovation in the wood products sector encompasses many themes, including new product development, the use of clusters to enhance competitiveness, and corporate 'thinking' to guide innovation. The idea of corporate social responsibility (CSR), and more broadly corporate culture, may ultimately guide innovative uses of timber and wood products, especially for products among the largest wood products and paper companies (Han et al., 2013). As consumers become more environmentally aware and increasingly have global procurement options, successful wood product firms will need innovative solutions to meet these demands. Ultimately, innovative social responsibility may guide corporate behavior more than profitability alone.

## 5 Meeting global demands for wood and bio-based products

Global timber markets are affected by many factors, including fiber supply, economic growth of timber consuming nations, and technical advances in harvesting, manufacture, and trade (Tromborg et al., 2000). Successful use of forest products will depend on accurate assessments of consumer demands, development of markets for specific timber-based products, and innovation to identify new and emerging opportunities. An effective global forest products market will need to address the dynamic international transportation and supply chains as well as recognizing green product attributes in demand by consumers (Hetemaki and Hurmekoski, 2016). Forests can sustainably and efficiently contribute to such intensified goals; however, changes may be needed for current practices including more efficient use of high-value timber resources as well as widespread cascading of wood products, leading to greater use of biomass energy products originally derived from timber products (Sikkema et al., 2017).

The global demand for forest products has grown steadily over the past 40 years; wood fiber and related international trade has grown at an even greater rate (Buongiorno et al., 2003). However, the US share of industrial roundwood production has decreased considerably over the past few decades, from 28% of global share in 1999 to about 17% in 2013 (Wear et al., 2015).

Four key drivers for global timber utilization have been identified by Bowyer (2006):

- globalization, with a shift in capacity to low-cost locations,
- fast-growing tree plantations to supply industrial wood,

- continued development of wood-based composites, and
- emerging markets in wood products (including China, eastern Europe, and Latin America).

Other global trends are already underway, including the role of climate change in influencing timber supplies, and the role of international trade of timber and bioenergy products in world markets. The social values of ecosystem services associated with timber products should not be overlooked. Many forestland resources other than timber (such as wetland values, agroforestry products, recreation, and watersheds) are playing a key role in guiding forest policy. Thus, sustainable forest management implies much more than simply maintaining a non-declining global supply of timber (Ince, 2010).

Lastly, changes in regional 'woodbaskets' will also influence global demand. Some (Toppinen et al., 2018; Jonsson, 2011) predict the likelihood of greater proportions of wood fiber being met from regions having higher net productivity (i.e. tropical forests). The global demand and the mix of products demanded will also be dynamic. The Global Forest Products Model (GFPM) was developed to forecast the demand, supply, prices, and trade of forest products to year 2060 and beyond. Under certain scenarios being explored by the International Panel on Climate Change (IPCC), a sixfold increase in demand for biofuels could result in parity between wood energy products and industrial roundwood. If this price convergence were to occur, industrial roundwood would see increased use for energy products and less for sawn lumber products; however, a side effect would be increased ecological stress in forests within certain countries (Raunikar et al., 2010). Thus, future global drivers for timber are likely to depend on a myriad of economic, social, and environmental factors. Sims (2003) describes the 'triple bottom line' of environmental, economic, and social benefits within the context of sustainable biomass utilization; however, the same principles apply to timber-based products. For wood product markets to be successful in the long term, producers and consumers must look beyond purely economic values. Thus, their full value must be included in any comparative assessment of timber-derived products and other alternatives.

Global timber markets have been examined from a number of perspectives which will be considered separately:

1 *Economic conditions and technological trends*
   Global timber markets are affected by many factors, including fiber supply, economic growth of timber consuming nations, and technical advances in harvesting, manufacture, and trade. Tromborg et al. (2000) adapted partial equilibrium models from the GFPM, considering 16 products within eight geographic regions. Their model incorporated both market optimization in the short run as well as imperfect foresight

of decision-makers to represent a longer-term component. They found that real prices for sawlogs were generally less volatile that those for pulpwood and wood waste residues.

2 *Future trends and developments*

Jonsson (2011) analyzed global trend wood product markets by considering two influential factors: (1) whether current patterns of globalization will continue or be supplanted by regionalism, and (2) whether social concern for environmental factors and climate change will manifest in strong policy measures.

Sikkema et al. (2017) suggested that an approach to sustainably meet EU wood could involve both the sourcing of high-quality wood fiber for energy and construction materials, as well as efficient cascaded use of materials. Innovative uses of wood materials and new classes of products are recognized as central elements of an emerging bioeconomy that could influence global timber demand. Winistorfer (2005) recognized the importance of global manufacturing and supply chains in satisfying major consumer markets, while recognizing the importance of specific industry segments and regions (e.g. furniture producers in the United States) to adapt to changing world conditions in order to remain competitive, even in light of declining industry infrastructure.

## 5.1 Timber products from plantation forests

On a global forest basis, plantations can play a key role in meeting wood fiber demands. For example, in 2000, forest plantations produced close to 27% of global industrial wood, yet occupied less than 3.5% of the forest land worldwide (Bowyer et al., 2005a; Brooks, 2001). Although forest plantations currently occupy only about 4% of global forest land area (FAO, 2006), plantations are expected to provide close to 50% of global roundwood production by year 2040 (Kanninen, 2010). Financial returns of plantation forests can vary from region to region, whether pulpwood is included, by international demands for wood products, and by the overall stability of nations (Cubbage et al., 2010). Species selection is also an important factor; for example, eucalyptus returns are often greater than those for pine species (Cubbage et al., 2010). As more and more plantation wood enters the supply stream, wood quality will continue to be an important concern. Regions such as the southeastern United States, that harvest wood on short rotations, can experience concerns regarding wood quality, such as the proportion of juvenile wood. Clark et al. (2006) found that juvenile wood periods are difficult to precisely determine, but that proximity to coastal regions could lead to shorter juvenile periods (sometimes less than 6 years), resulting in overall greater wood quality. Globally, the impact of increased proportions of juvenile wood is also a concern, yet could accelerate the growth of the wood composites industry, where the presence of juvenile wood is less of a detriment (Bowyer, 2001).

Advantages of forest plantations include the establishment of productive forests from relatively small land areas, and reduced pressure for harvests from natural forests. However, there are also concerns regarding ongoing use of plantation forests. Among them are the potential loss of soil fertility and productivity under short rotations, increased risk of disease and insect infestations due to cultivation of monocultures, reduced biological diversity, and potential invasive or exotic species introductions (Bowyer et al., 2005a). These concerns can be ameliorated through prudent plantation management practices including matching species to sites, use of genetically diverse planting stock, and careful monitoring of plantations.

On an international basis, the financial returns of plantations also must account for nation-specific risk factors. Factors that could adversely influence plantation businesses include political risks, changes in commercial conditions and/or currency exchange rates, war, or government actions and policies—all of which could influence the ease of doing business (Cubbage et al., 2010). This would also extend to plantations of exotic species, including genetically engineered eucalyptus plantations in the United States (Wear et al., 2015). However, complex valuations and returns could be more difficult to evaluate and therefore more risky for investors, especially when considering future markets for cellulose, especially for bioenergy feedstock.

### 5.2 Ecosystem services

Well-established financial metrics (such as internal rate of return, NPV, and payback period) often do not consider important ecosystem services—especially for natural stands. These services could include water benefits, forage, preservation of wetlands and/or estuaries, carbon benefits, alternative land uses, and potential wildfire risk reductions. The role of ecosystem services could be pivotal in the aggregate demand for timber products. Millennium Ecosystem Assessment (MEA, 2005) recognizes four categories including provisioning, regulating, supporting, and cultural services. Ecosystem services can also include secondary benefits such as reducing the risk of wildfires, increasing carbon sequestration from forests, and use of biomass as an alternative to burning fossil fuels for energy (Oliver and Deal, 2007). Other key services include air quality improvements if bioenergy is generated in place of fossil energy and potential water-quality improvements (especially after a wildfire occurrence) (Deal et al., 2012). Thus, the four categories of ecosystem services together provide a diverse array of services and benefits to society.

In the EU, Jonsson (2013) found that European forest resources may be insufficient to meet future demands for wood fiber, resulting in trade-offs and prioritization of various ecosystem services. The full range of ecosystem services will also need to be evaluated in this context to assess market and non-market conditions. Current forest management conditions in the EU favor valuing a broad array of forest ecosystem services to determine total benefits (Jonsson, 2013).

## 6 Customization of timber-derived products in an era of globalization[1]

The timber industry must remain competitive to meet future global challenges of supplying wood fiber and products to a growing population. A central part of competitive advantage will be meeting consumer needs with high-quality timber-derived products in an agile and responsive, or 'customizable' manner. Mass customization has been defined as 'production of individually customized products using flexible and responsive manufacturing systems at a cost and speed near that of mass-produced items' (Stump and Badurdeen, 2012; Azouzi et al., 2009a). Customization has gained prominence in a number of manufacturing sectors since the late 1980s (Fogliatto et al., 2012), and notably within the wood products sector during the last decade (Lihra et al., 2008; Nicholls and Bumgardner, 2018).

Product customization is a pathway for producers to be responsive to consumers by providing increased options for product choices, styles, and delivery schedules among other factors. Of importance to timber and solid wood products is that customization represents a paradigm shift away from the production of commodity products such as lumber toward products with more specific applications. Product customization has most recently been implemented by much of the North American wood products industry, and this was at least partially in response to the recession of 2008-2009, and in an effort to mitigate its effects. Customization also encompasses many other elements such as global competition, agility, lean manufacturing, and clustering, addressed later in this section.

The US hardwood lumber industry provides a representative case study for how mass customization is transforming this segment of the timber industry. Hardwood producers face new challenges in dynamic international markets, and this sector now must compete with global supplies of raw materials and global manufacturers to meet the needs of US consumers. The recent loss of manufacturing infrastructure for some components of the North American industry has brought high levels of change and uncertainty. Of concern is the ability to competitively manufacture and maintain supply chains, while providing stability to other sectors of the overall forest industry (Winistorfer, 2005). In addition, repercussions of the housing crisis are still being felt today across

1 Much of this material has been derived from Nicholls and Bumgardner (2018).

Published by Burleigh Dodds Science Publishing Limited, 2020.

many timber-related sectors, which has created new ways of thinking for firms to do business and interact within their supply chains. Thus, current economic events can strongly influence customization strategies by lumber producers as well as the markets they serve, which are increasingly global in nature.

For wood product industries to realize competitive and comparative advantages, they must identify their strengths at all stages within the value chain. This was especially evident during the economic crisis of the last decade. Panwar et al. (2012) considered the macroeconomic factors and resulting dislocations felt by domestic wood products firms during the 2008 recession, identifying specific actions that could help maintain competitive advantage throughout the forest sector supply chain during periods of economic crisis. Firms that successfully adapted to decreased demands, fluctuating prices and exchange rates, and greater competition were better able to survive the recession (by realizing a comparative advantage). They also identified strategies for maintaining coherent and intact supply chains during times of highly volatile markets (by realizing a competitive advantage).

Optimum strategies for maintaining and enhancing competitive advantage are constantly evolving; strategies that have worked well in recent business cycles may not be as effective in the future. However, it is clear that as the operating environment of the forest sector becomes more complex, maintaining competitiveness will also become increasingly complicated (Korhonen et al., 2018). Brown and Blackmon (2005) introduced strategic approaches for remaining competitive during economic volatility. They discussed two ways that firms can gain competitive advantage. In market-led views, external opportunities in all markets should be identified; in resource-based views, firms should mobilize resources that are likely to maximize returns. Successful strategies are then based on approaches that consider both market requirements and manufacturing capabilities under dynamic and unpredictable business environments. Of importance to timber-related products is that new production paradigms (such as mass customization) offer increased flexibility for secondary manufacturers. This can influence all stages of the supply chain in that efficient wood fiber use here will allow for more flexibility in production of other wood-based products from limited timber resources.

In North America, Europe, and some parts of Asia, mass customization, lean manufacturing, strategic supply chains, and other strategies for enhancing competitiveness have been explored (Jiao et al., 2003). In the US wood products industry, a paradigm shift is being experienced in sectors strongly influenced by globalization. One example is the wood household furniture industry (Song and Gazo, 2013; Schuler and Buehlmann, 2003), where customized production and strategic supply chain alliances are key elements of competitiveness.

A good example of customization is within the Amish furniture industry in the northeastern United States. Here, consumers are involved in product assembly decisions. Multiple choices in species, finish, and hardware are offered at the retail level and consumer orders are then placed with local manufacturers. Consumers can even specify whether to use character-marked or naturally blemished woods (Nicholls and Bumgardner, 2018). Dugan (2009) has called for 'new rules' to guide the US furniture industry, including agility, niche marketing, supply chain development, and lean production. Azouzi et al. (2009b) claim that innovative priorities for furniture industries also include short lead times, high product quality, product variety, and a focus on profit margin.

The marketing of timber-derived secondary products must also be adaptable to changing conditions. A key component of this paradigm shift is the desire of consumers for customized products (Pine, 1993). Shortening lead times associated with customization is central to competitiveness, and can be enhanced through lean manufacturing, supply chain agility, and local sourcing and purchasing practices. Product modularity, competitive cost structures, and integrated supply chains have also been identified as critical elements of a furniture customization structure (Buehlmann, 2009). Gilmore and Pine (1997) suggest that consumers are increasingly moving away from a mindset of standardization simply to achieve a lower price, thereby creating 'customer sacrifice gaps'. Thus, producers in higher-cost locations can compete by focusing less on price and more on customization of goods (Mishra et al., 2014). Ultimately, customization may include scenarios in which end-users are able to design products prior to purchase. This flexibility in marketing to consumers could very well determine which firms remain competitive and profitable. A related trend is the 'greening' of the supply chain. Green supply chain management has been found to influence environmental, operational, and economic performance (Nicholls and Bumgardner, 2018). The most important link in the supply chain – consumers – will be the ultimate determinant of how widely sustainability measures and marketing methods are accepted.

Successful mass customization in timber-derived products will have broad-reaching economic effects throughout the forest value chain. This includes not only manufacturing, but also product sourcing, procurement, silviculture, harvesting, and transportation. All of these can influence wood quality and product quality.

Consumer preferences for the next generation of timber-based products are likely to be influenced by many factors including customization, competitiveness, innovation, resilience, sustainability, global timber demand among others. These elements will ultimately guide the mix of successful products that emerge within the marketplace. Increased customization of wood products will be needed to meet consumers' increasing demands for product variety. Customization is relevant for timber products given the importance of staying current with consumer markets during dynamic and challenging business cycles. Greater customization by companies requires a combination of technology investment, training, and management insight to be successful. Product differentiation and customization will also play a role in competitiveness at the firm level as timber products become more integral within the bioeconomy, and firms realize the need to stay agile and responsive throughout the value chain. The competitive dynamics between firms within an industry will gain importance given the potential substitution of raw materials within the bioeconomy, and the need to find innovative uses and markets (Korhonen et al., 2018).

A key link needed for successful customization is the supply chain, where companies can leverage both internal and external factors to enhance their competitiveness through agility and customization. Um (2017) found that improvements in supply chain agility influenced customer service and differentiation positively, with less overall benefit to business performance and profitability. Many of the supply chain efficiencies can be found in economic clusters that have become well established. As firms become more competitive and more complex, they may enter into a phase known as 'coopetition' (a hybrid of cooperation and competition) that allows them to function as mature clusters with well-established supply chains (Hartono and Sobari, 2016). Coopetition also can enhance communication and knowledge exchange among producers and customers.

Equally important within the context of customization is the need for timber-related firms to find international opportunities can lead to geographically diverse value chains. For example, Mohuiddan and Su (2014) suggest that, rather than a 'local vs. foreign' mindset, successful firms will be able to integrate both locally produced and imported materials. Building 'relationship value' with customers has also been identified as a key component of wood products firm success—especially in the context of customized products, where strategic relationships with consumers can lead to opportunities that are driven by more than price (Lefaix-Durand et al., 2009). Further, as markets tend toward a greater emphasis on 'green' products, the sustainability aspects of these products will become an integral aspect of customization and competitiveness (Carvalho et al., 2010; Chavez et al., 2016; Green et al., 2012).

## 7 Case study: Scandinavian practices in the timber industry and forest sector

Diverse themes related to the development of forest products, timber markets, and quality improvements have been considered in this book chapter. An overarching theme of these topics has been sustainability, environmental awareness, and GHG reductions for climate change mitigation. In Scandinavian nations, including Sweden and Finland, the timber products value chain has been highly developed and serves as a representative case study. This includes wood procurements, harvesting, transportation, manufacturing, energy generation, and use. Sustainability initiatives are well developed within the EU, and particularly within the Finnish forest products industry (Husgafvel et al., 2013).

Bolkesjo et al. (2006) used partial equilibrium models to evaluate three energy price scenarios in Norway, finding that production levels in most forest industry sectors were relatively unaffected by moderate changes in energy prices. Scandinavia could be considered a world leader not only in forest operations and timber products development, but also in climate change mitigation and carbon sequestration by wood products.

Over the past several decades, advanced harvesting methods and whole-tree utilization systems have been well developed in Scandinavia. At the same time, new applications for timber products, such as all-wood multistory construction, cascaded use of wood products, and more accurate quantification of carbon pools and benefits, have taken place. Undergirding these developments has been national policies favoring the use of renewable materials instead of fossil-based resources. For example, the development of efficient fluidized bed combustion systems has become a leading wood energy heat and power production tool in Finland. This case study highlights how sustainability measures can be adopted to value both carbon sequestration and wood products development through successful timber management practices.

Several elements of Scandinavia's success are worth noting. Perhaps most significant is the numerous policy measures designed to encourage use of renewable resources. Sweden has implemented a carbon tax (started in 1991) that has spurred the use of wood gasification systems, bioenergy systems, and numerous other avenues for substituting biomass products in place of fossil fuels (Scharin and Wallstrom, 2018). Forest biorefineries have also been identified as a viable path forward for Scandinavia, given their extensive forest resources, well-developed supply chains, efficient harvest and transportation networks, and pulp and paper infrastructure (Hamalainen et al., 2011). Although biorefineries often are centered on the production of liquid fuels, solid wood and timber-derived products are still an integral part of the supply and product chain (Larsson et al., 2016).

More recently, timber structures have become more mainstream, and high-rise applications have been identified as a potential climate change mitigation measure that represents opportunities for carbon sequestration (Skullestad et al., 2016). Here, timber structures were found to create climate change impacts of up to 84% lower than for similar reinforced concrete structures (based on $CO_2$-equivalents per square meter of floor area). Eriksson et al. (2012) also documented climate change mitigation benefits through increased wood use in the European construction sector. From an operational standpoint, the complex carbon flows and pools have been modeled in Norway in forest optimization models (Raymer et al., 2009). Here, overall timber value was generally reduced under scenarios of maximizing carbon benefit; however, if the substitution benefits of replacing energy-intensive materials with wood are considered, NPV of forest stands can be increased by roughly a factor of 1.5.

Broad-based and overarching interactions between timber, forest products, and bioenergy products influence how wood fiber is used, including the integration of supply chains with resources and products. The future path of commodity energy prices is a wild-card factor which could influence the balance between products not only in Scandinavia but worldwide. Solid wood (commodity) products typically have more price stability than energy products. Innovative use of wood resources in Scandinavia extends to supply chain management (from forest to mill) as well as manufacturing practices. Policy measures, forest industry infrastructure, harvesting and in-woods innovations, pulp and paper facilities that have been the backbone of the wood-using industries, and forest clusters have lead to increased efficiencies.

## 8 Future trends

The twenty-first century can be seen as a turning point for global forest product markets. Some (Hetemaki and Hurmekoski, 2016) have characterized its current status as being in a phase of 'creative destruction'. That is, an era characterized by major declines in some timber-based sectors, with a simultaneous emergence of new products and businesses. Challenging questions that will need to be addressed will likely include:

- how will timber/solid wood products fit into the bioeconomy?
- what level of innovation will be needed to compete globally?
- what level of industry and product diversification will be needed to compete in international markets within acceptable levels of risk?

Successful timber-related industries will need to develop sufficient tools to be responsive to economic downturns, to adapt with agility to changing conditions, and to manage risk within acceptable parameters. Certain themes will remain important including the role of government policies, international trade, and competition. Future approaches will need to consider equally important issues, such as value-added processing, wood-sector employment, structural changes within wood producing industries, the diffusion of new products and services, and the role of sustainability (Hetemaki and Hurmekoski, 2016). LCA is a valuable tool to quantify sustainability aspects; however, it is only as good as the underlying data, often provided empirically. Future forest sector status will likely also be in a position of valuing ecosystem services based on their full societal values rather than just their intrinsic economic values. Product customization is also seen as a major opportunity for producers to meet consumer demands

Published by Burleigh Dodds Science Publishing Limited, 2020.

more efficiently, and this applies to all stages of the product value chain (from woods to mill to consumer).

As the bioeconomy progresses in the coming decades, and supply chains become fully developed, it is anticipated that an economic mix of timber, solid wood, and liquid fuel products will become integrated at appropriate scales. As more and more nations adopt more stringent climate change mitigation measures, the role of the bioeconomy will become more significant; however, it remains to be seen which products (e.g. construction materials versus transportation fuels) will become more prevalent. LCA, which will quantify carbon flows, energy balances, and material utilization, will play an integral role in the future bioeconomy, in conjunction with sustainable policies.

## 9 Abbreviations

CLT       cross-laminated timber
$CO_2$    carbon dioxide
EU        European Union
FAO       Food and Agricultural Organization
GHG       greenhouse gas
GFPM      Global forest products model
IPCC      International Panel on Climate Change
LCA       life cycle assessment
NPV       net present value
Pg        Petagram
Tg        Terragram
UN        United Nations
UK        United Kingdom

## 10 Where to look for further information

Numerous sources are available for additional information regarding future developments in markets for wood timber products. Included is a short list of some of the most informative references:

- Bergman, R., Puettmann, M., Taylor, A. and Skog, K. E. 2014. The carbon impacts of wood products. *Forest Products Journal* 64(7-8), 220-31.
- Bowyer, J. L., Bratkovich, S., Howe, J., Fernholz, K., Frank, M., Hanessian, S., Groot, H. and Pepke, E. 2016. *Modern Tall Wood Buildings: Opportunities for Innovation*. Dovetail Partners, Minneapolis, MN. 16p. Available at: http://www.dovetailinc.org/report_pdfs/2016/dovetailtallwoodbuildings0116.pdf.

- Bugge, M. M, Hansen, T. and Klitkou, A. 2016. What is the bioeconomy? A review of the literature. *Sustainability* 8(7), 691.
- Eriksson, L. O., Gustavsson, L., Hänninen, R., Kallio. M., Lyhykäinen, H., Pingoud, K., Pohjola, J., Sathre, R., Solberg, B., Svanaes, J. and Valsta, L. 2012. Climate change mitigation through increased wood use in the European construction sector–towards an integrated modelling framework. *European Journal of Forest Research* 131, 131–44.
- Hansen, E. N. 2010. The role of innovation in the forest products industry. *Journal of Forestry* 108(7), 348–53.
- Kanninen, M. 2010. Plantation forests: global perspectives. In: Bauhus, B., van der Meer, P. and Kanninen, M. (Eds), *Ecosystem Goods and Services from Plantation Forests* (1st edn.). London, 272p. doi:10.4324/9781849776417.
- Lippke, B., Wilson, J., Meil, J. and Taylor, A. 2010. Characterizing the importance of carbon stored in wood products. *Wood and Fiber Science* 42, 5-14.
- Nicholls, D. L. and Bumgardner, M. S. 2018. Challenges and opportunities for North American hardwood manufacturers to adopt customization strategies in an era of increased competition. *Forests* 9(4), 186. doi:10.3390/f9040186.
- Pingoud, K., Pohjola, J. and Valsta, L. 2010. Assessing the integrated climatic impacts of forestry and wood products. *Silva Fennica* 44(1), 155-75.
- Sathre, R. and Gustavsson, L. 2006. Energy and carbon balances of wood cascade chains. *Resources, Conservation and Recycling* (47), 332-55.

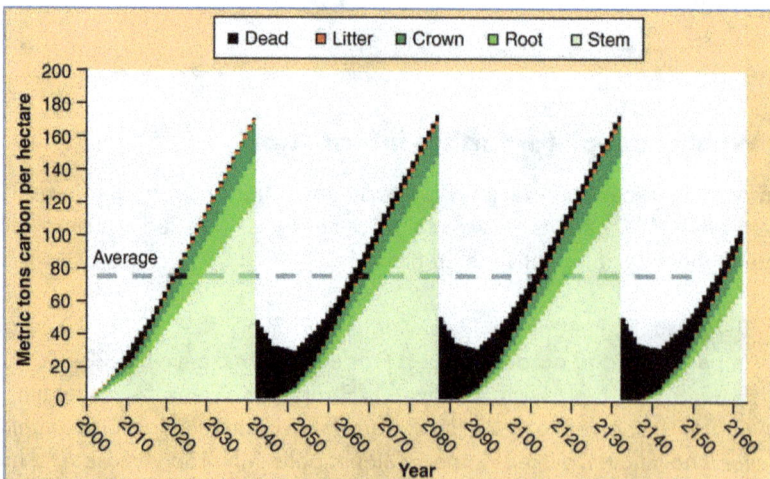

Source: Lippke et al. (2010).

Published by Burleigh Dodds Science Publishing Limited, 2020.

**Table 1** Carbon implications of increased use of wood in US construction

| Type of building | Approximate additional wood volume annually (billion board feet) | Additional annual carbon benefit (million metric tons $CO_2e$) | Equivalent number of passenger vehicles off the road for a year | Equivalent number of coal fired power plants shut down for a year |
|---|---|---|---|---|
| Low-rise nonresidential | 4.5 | 19 | 4 100 000 | 5 |
| Multifamily | 0.7 | 3 | 700 000 | 1 |
| US buildings 7- to 15-stories | 1.6-2.4 | 7-10 | 1 500 000-2 200 000 | 2-3 |
| Aggregate | 6.8-7.6 | 29-33 | 6 300 000-7 000 000 | 8-9 |

*Source:* Bowyer et al. (2016).

Published by Burleigh Dodds Science Publishing Limited, 2020.

**Carbon displaced/carbon in wood used**

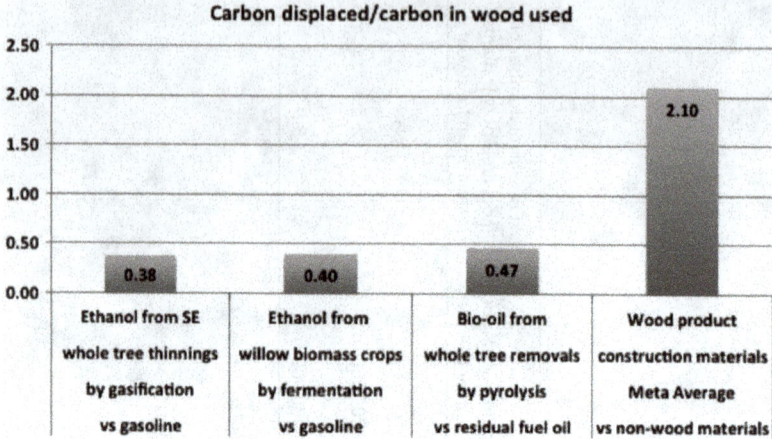

| | | | |
|---|---|---|---|
| 0.38 | 0.40 | 0.47 | 2.10 |
| Ethanol from SE whole tree thinnings by gasification vs gasoline | Ethanol from willow biomass crops by fermentation vs gasoline | Bio-oil from whole tree removals by pyrolysis vs residual fuel oil | Wood product construction materials Meta Average vs non-wood materials |

Source: Lippke et al. (2012).

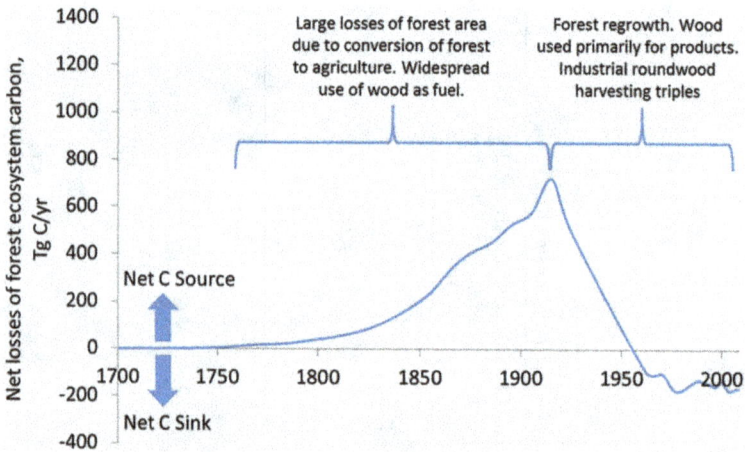

Source: Malmsheimer et al. (2011).

# 11 References

Azouzi, R., Beauregard, R. and D'Amours, S. 2009a. Exploratory case studies on manufacturing agility in the furniture industry. *Management Research News* 32(5), 424–39. doi:10.1108/01409170910952930.

Azouzi, R., D'Amours, S. and Beauregard, R. 2009b. An agility reference model for the manufacturing enterprise: the example of the furniture industry. In: Piller, F. T. and Tseng, M. M. (Eds), *Handbook of Research in Mass Customization and Personalization*. World Scientific Publishing Co. Pvt Ltd., pp. 403–26.

Bais-Moleman, A. L., Sikkema, R., Vis, M., Reumerman, P., Theurl, M. C. and Erb, K.-H. 2018. Assessing wood use efficiency and greenhouse gas emissions of wood product cascading in the European Union. *Journal of Cleaner Production* 172, 3942–54. doi:10.1016/j.jclepro.2017.04.153.

Batsy, D. R., Solvason, C. C., Sammons, N. E., Chambost, V., Bilhartz II, D. L., Eden, M. R., El-Halwagi, M. M. and Stuart, P. R. 2013. Product portfolio selection and process design for the forest biorefinery. In: Stuart, P. R. and El-Halwagi, M. M. (Eds), *Integrated Biorefineries-Design, Analysis, and Optimization*. Taylor & Francis Group, Abingdon, U.K.

Bolkesjo, T. F., Erik Tromborg, E. and Solberg, B. 2006. Bioenergy from the forest sector: economic potential and interactions with timber and forest products markets in Norway. *Scandinavian Journal of Forest Research* 21(2), 175–85.

Bowyer, J. L. 2006. Changing realities in forest sector markets. Food and Agriculture Organization. Available at: http://www.fao.org/3/y5918e/y5918e13.htm#TopO fPage.

Bowyer, J. L., Howe, J., Guillery, P. and Fernholz, K. 2005a. Fast-growth tree plantations for wood production – environmental threat or a means of "saving" natural forests? Dovetail Partners, Inc., 16p. Available at: http://www.dovetailinc.org/report_pdfs/2 005/dovetailplant1005b.pdf.

Bowyer, J. L., Briggs, D., Lippke, B., Perez-Garcia, J. and Wilson, J. 2005b. Life cycle environmental performance of renewable materials in the context of residential building construction. Phase I research report. CORRIM, Seattle, WA.

Bowyer, J. L., Bratkovich, S., Howe, J., Fernholz, K., Frank, M., Hanessian, S., Groot, H. and Pepke, E. 2016. Modern tall wood buildings: opportunities for innovation. Dovetail Partners, Minneapolis, MN, 16p. Available at: http://www.dovetailinc.org/report_p dfs/2016/dovetailtallwoodbuildings0116.pdf.

Brooks, D. A. 2001. Global and regional demand for forest products, now and in the future. In: *Proceedings: Biotech Branches Out – a Look at the Opportunities and Impacts of Forest Biotechnology*. Pew Initiative on Food and Biotechnology, Atlanta, 4–5 December. Available at: http://pewagbiotech.org/events/1204/presentations/ BrooksEdited.ppt#1.

Brown, S. and Blackmon, K. 2005. Aligning manufacturing strategy and business-level competitive strategy in new competitive environments: the case for strategic resonance. *Journal of Management Studies* 42(4), 793–815. doi:10.1111/j.1467-6486.2005.00519.x.

Buehlmann, U. 2009. Opportunities and challenges of furniture manufacturers implementing mass customization. In: Piller, F. T. and Tseng, M. M. (Eds), *Handbook of Research in Mass Customization and Personalization* (in 2 Vols.). World Scientific. Available at: https://ssrn.com/abstract=1596414.

Bugge, M. M., Hansen, T. and Klitkou, A. 2016. What is the bioeconomy? A review of the literature. *Sustainability* 8(7), 691. doi:10.3390/su8070691.

Buongiorno, J., Zhu, S., Zhang, D., Turner, J. and Tomberlin, D. 2003. *The Global Forest Products Model (GFPM): Structure, Estimation, Applications*. Elsevier Science, 301p.

Carvalho, H., Azevedo, S. G. and Cruz-Machado, V. 2010. Supply chain performance management: lean and green paradigms. *International Journal of Business Performance and Supply Chain Modelling* 2, 303–33.

Chavez, R., Yu, W., Feng, M. and Wiengarten, F. 2016. The effect of customer-centric green supply chain management on operational performance and customer satisfaction. *Business Strategy and the Environment* 25(3), 205–20. doi:10.1002/bse.1868.

Ciccarese, L., Pellegrino, P. and Pettenella, D. 2014. A new principle of the European Union Forest Policy: the cascading use of wood products. *Italian Journal of Forest and Mountain Environments* 69(5), 285–90. doi:10.4129/ifm.2014.5.01.

Clark, A., Daniels, R. F. and Jordan, L. 2006. Juvenile/mature wood transition in loblolly pine as defined by annual ring specific gravity, proportion of latewood, and microfibril angle. *Wood and Fiber Science* 38(2), 292–9.

Crampton, A. 2017. Cross-laminated timber: an innovative building material takes hold in Oregon. *Metroscape*, Winter 2017, pp. 6–12.

Crespell, P., Knowles, C. and Hansen, E. 2006. Innovativeness in the North American softwood sawmilling industry. *Forest Science* 52(5), 568–78.

Cubbage, F., Koesbandana, S., Mac Donagh, P., Rubilar, R., Balmelli, G., Olmos, V. M., De La Torre, R., Murara, M., Hoeflich, V. A., Kotze, H., Gonzalez, R., Carrero, O., Frey, G., Adams, T., Turner, J., Lord, R., Huang, J., MacIntyre, C., McGinley, K., Abt, R. and Phillips, R. 2010. Global timber investments, wood costs, regulation, and risk. *Biomass and Bioenergy* 34(12), 1667–78. doi:10.1016/j.biombioe.2010.05.008.

Damaspatra, S. 2009. Future market drivers for the forest products industry. *BioResources* 4(4), 1263–6.

Deal, R. L., Cochran, B. and Larocco, G. 2012. Bundling of ecosystem services to increase forestland value and enhance sustainable forest management. *Forest Policy and Economics* 17, 69–76. doi:10.1016/j.forpol.2011.12.007.

Dugan, M. K. 2009. *The Furniture Wars: How America Lost a Fifty Billion Dollar Industry*. Goosepen Press, Conover, NC, 450p. Available at: https://www.amazon.com/Furnitu re-Wars-Michael-K-Dugan-ebook/dp/B006MXATLO/ref=dp_kinw:strp_1/141-670 0370-5153057.

Elia, J. A. and Floudas, C. A. 2014. Energy supply chain optimization of hybrid feedstock processes: a review. *Annual Review of Chemical and Biomolecular Engineering* 5, 147–79. doi:10.1146/annurev-chembioeng-060713-040425.

Eriksson, L. O., Gustavsson, L., Hänninen, R., Kallio, M., Lyhykäinen, H., Pingoud, K., Pohjola, J., Sathre, R., Solberg, B., Svanaes, J. and Valsta, L. 2012. Climate change mitigation through increased wood use in the European construction sector—towards an integrated modelling framework. *European Journal of Forest Research* 131(1), 131–44. doi:10.1007/s10342-010-0463-3.

Fast, P., Gafner, B., Jackson, R. and Li, J. 2016. Case study: an 18 storey tall mass timber hybrid student residence at the University of British Columpia, Vancouver. *Proceedings of World Conference on Timber Engineering*, Vienna, Austria, 9p.

Fogliatto, F. S., da Silveira, G. J. C. and Borenstein, D. 2012. The mass customization decade: an updated review of the literature. *International Journal of Production Economics* 138(1), 14–25. doi:10.1016/j.ijpe.2012.03.002.

Food and Agriculture Organization. 2006. *Global Forest Resources Assessment 2005*. FAO, Rome, Italy.

Garrity, D. P. 2004. Agroforestry and the achievement of the Millennium Development Goals. *Agroforestry Systems* 61, 5–17.

Gilmore, J. H. and Pine, B. J. 1997. The four faces of mass customization. *Harvard Business Review* 75(1), 91–101.

Green Jr., K. W., , Zelbst, P. J., Meacham, J. and Bhadauria, V. S. 2012. Green supply chain management practices: impact on performance. *Supply Chain Management: an International Journal* 17(3), 290-305. doi:10.1108/13598541211227126.

Gustavsson, L., Joelsson, A. and Sathre, R. 2010. Life cycle primary energy use and carbon emission of an eight-storey wood-framed apartment building. *Energy and Buildings* 42(2), 230-42. doi:10.1016/j.enbuild.2009.08.018.

Hämäläinen, S., Näyhä, A. and Pesonen, H. L. 2011. Forest biorefineries–a business opportunity for the Finnish forest cluster. *Journal of Cleaner Production* 19(16), 1884-91. doi:10.1016/j.jclepro.2011.01.011.

Han, X., Hansen, E., Panwar, R., Hamner, R. and Orozco, N. 2013. Connecting market orientation, learning orientation and corporate social responsibility implementation: is innovativeness a mediator? *Scandinavian Journal of Forest Research* 28(8), 784-96. doi:10.1080/02827581.2013.833290.

Hansen, E. N. 2010. The role of innovation in the forest products industry. *Journal of Forestry* 108(7), 348-53.

Hansen, E. 2014. Innovativeness in the face of decline: performance implications. *International Journal of Innovation Management* 18(5). doi:10.1142/S136391961450039X.

Hansen, E., Juslin, H. and Knowles, C. 2007. Innovativeness in the global forest products industry: exploring new insights. *Canadian Journal of Forest Research* 37(8), 1324-35. doi:10.1139/X06-323.

Hansen, E., Nybakk, E. and Panwar, R. 2014. Innovation insights from North American forest sector research: a literature review. *Forests* 5(6), 1341-55. doi:10.3390/f5061341.

Hartono, S. and Sobari, A. 2016. Coopetition, cluster externalities and company performances: formation for competitiveness of the wood and rattan furniture industry. *International Journal of Organizational Innovating* 9(2), 271-86.

Hennigar, C. R., MacLean, D. A. and Amos-Binks, L. J. 2008. A novel approach to optimize management strategies for carbon stored in both forests and wood products. *Forest Ecology and Management* 256(4), 786-97. doi:10.1016/j.foreco.2008.05.037.

Hetemäki, L. and Hurmekoski, E. 2016. Forest products markets under change: review and research implications. *Current Forestry Reports* 2(3), 177-88. doi:10.1007/s40725-016-0042-z.

Hildebrandt, J., Hagemann, N. and Thrän, D. 2017. The contribution of wood-based construction materials for leveraging a low carbon building sector in Europe. *Sustainable Cities and Society* 34, 405-18. doi:10.1016/j.scs.2017.06.013.

Hoglmeier, K., Steubing, B., Weber-Blaschke, G. and Richter, K. 2015. LCA-based optimization of wood utilization under special consideration of a cascading use of wood. *Journal of Environmental Management* 152, 158-70. doi:10.1016/j.jenvman.2015.01.018.

Hong, B. H., How, B. S. and Lam, H. L. 2016. Overview of sustainable biomass supply chain: from concept to modelling. *Clean Technologies and Environmental Policy* 18(7), 2173-94. doi:10.1007/s10098-016-1155-6.

Hovgaard, A. and Hansen, E. 2004. Innovativeness in the forest products industry. *Forest Products Journal* 54(1), 26-33.

Hurmekoski, E., Jonsson, R. and Nord, T. 2015. Context, drivers, and future potential for wood-frame multi-story construction in Europe. *Technological Forecasting and Social Change* 99, 181-96. doi:10.1016/j.techfore.2015.07.002.

Husgafvel, R., Watkins, G., Linkosalmi, L. and Dahl, O. 2013. Review of sustainability management initiatives within Finnish forest products industry companies–translating EU level steering into proactive initiatives. *Resources, Conservation and Recycling* 76, 1–11. doi:10.1016/j.resconrec.2013.04.006.

Ince, P. J. 2010. Global sustainable timber supply and demand. In: *Sustainable Development in the Forest Products Industry*. Universidade Fernando Pessoa, Porto, Portugal, pp. 29–41. Chapter 2. ISBN: 9789896430528.

Janssen, M., Chambost, V. and Stuart, P. R. 2008. Successful partnerships for the forest biorefinery. *Industrial Biotechnology* 4(4), 352–62. doi:10.1089/ind.2008.4.352.

Jiao, J., Ma, Q. and Tseng, M. M. 2003. Towards high value-added products and services: mass customization and beyond. *Technovation* 23(10), 809–21. doi:10.1016/S0166-4972(02)00023-8.

Jonsson, R. 2011. Trends and possible future developments in global forest-product markets–implications for the Swedish forest sector. *Forests* 2(1), 147–67. doi:10.3390/f2010147.

Jonsson, R. 2013. How to cope with changing demand conditions – the Swedish forest sector as a case study: an analysis of major drivers of change in the use of wood resources. *Canadian Journal of Forest Research* 43(4), 405–18. doi:10.1139/cjfr-2012-0139.

Kanninen, M. 2010. Plantation forests: global perspectives. In: Bauhus, B., van der Meer, P. and Kanninen, M. (Eds), *Ecosystem Goods and Services Rom Plantation Forests* (1st edn.). Earthscan Ltd, London, UK and Washington DC, 272p. doi:10.4324/9781849776417.

Kaziolas, D. N., Bekas, G. K., Zygomalas, I. and Stavroulakis, G. E. 2015. Life cycle analysis and optimization of a timber building. *Energy Procedia* 83, 41–9. doi:10.1016/j.egypro.2015.12.194.

Kircher, M. 2014. The emerging bioeconomy: industrial drivers, global impact, and international strategies. *Industrial Biotechnology* 10(1), 11–8. doi:10.1089/ind.2014.1500.

Korhonen, J., Hurmekoski, E., Hansen, E. and Toppinen, A. 2018. Firm-level competitiveness in the forest industries: review and research implications in the context of bioeconomy strategies. *Canadian Journal of Forest Research* 48(2), 141–52. doi:10.1139/cjfr-2017-0219.

Kramer, M. R. and Porter, M. 2011. Creating shared value. *Harvard Business Review* 89(1/2), 62–77.

Kubeczko, K., Rametsteiner, E. and Weiss, G. 2006. The role of sectoral and regional innovation systems in supporting innovations in forestry. *Forest Policy and Economics* 8(7), 704–15. doi:10.1016/j.forpol.2005.06.011.

Larsson, M., Stendahl, M. and Roos, A. 2016. Supply chain management in the Swedish wood products industry - a need analysis. *Scandinavian Journal of Forest Research* 31(8), 777–87. doi:10.1080/02827581.2016.1170874.

Lefaix-Durand, A., Kozak, R., Beauregard, R. and Poulin, D. 2009. Extending relationship value: observations from a case study of the Canadian structural wood products industry. *Journal of Business and Industrial Marketing* 24(5/6), 389–407. doi:10.1108/08858620910966273.

Lihra, T., Buehlmann, U. and Beauregard, R. 2008. Mass customization of wood furniture as a competitive strategy. *International Journal of Mass Customisation* 2(3/4), 200–14. doi:10.1504/IJMASSC.2008.017140.

Published by Burleigh Dodds Science Publishing Limited, 2020.

Lippke, B., Wilson, J., Meil, J. and Taylor, A. 2010. Characterizing the importance of carbon stored in wood products. *Wood and Fiber Science* 42, 5–14.

Lippke, B., Gustafson, R., Venditti, R., Steele, P., Volk, T. A., Oneil, E., Johnson, L., Puettmann, M. E. and Skog, K. 2012. Comparing life-cycle carbon and energy impacts for biofuel, wood product, and forest management alternatives. *Forest Products Journal* 62(4), 247–57. doi:10.13073/FPJ-D-12-00017.1.

Malmsheimer, R. W., Bowyer, J. L., Fried, J. S., Gee, E., Izlar, Robert, L., Miner, R. A., Munn, I. A., Oneil, E. and Stewart, W. C. 2011. Managing forests because carbon matters: integrating energy, products, and land management policy. *Journal of Forestry* 109(7S), S7–S50.

Margareta, R.-T. 2014. Clusters: innovation, knowledge and competitiveness in the wood processing industry . *SEA – Practical Application of Science* 6, 81–6.

Millennium Ecosystem Assessment. 2005. Available at: https://www.millenniumassessme nt.org/en/index.html.

Mishra, R., Pundir, A. K. and Ganapathy, L. 2014. Manufacturing flexibility research: a review of literature and agenda for future research. *Global Journal of Flexible Systems Management* 15(2), 101–12. doi:10.1007/s40171-013-0057-2. Flex, J. *Systems Management* 15, 101–12.

Mohiuddin, M. and Su, Z. 2014. Global value chains and the competitiveness of Canadian manufacturing SMEs. *Academy of Taiwan Business Management Review* 10(2), 82–92.

Nicholls, D. L. and Bumgardner, M. S. 2018. Challenges and opportunities for North American hardwood manufacturers to adopt customization strategies in an era of increased competition. *Forests* 9(4), 186. 16 p. doi:10.3390/f9040186.

Oliver, C. D. and Deal, R. L. 2007. A working definition of sustainable forestry and means of achieving it at different spatial scales. *Journal of Sustainable Forestry* 24(2–3), 141–63. doi:10.1300/J091v24n02_03.

Panwar, R., Vlosky, R. and Hansen, E. 2012. Gaining competitive advantage in the new normal. *Forest Products Journal* 62(6), 420–8. doi:10.13073/FPJ-D-12-00104.1.

Parker, N., Tittmann, P., Hart, Q., Nelson, R., Skog, K., Schmidt, A., Gray, E. and Jenkins, B. 2010. Development of a biorefinery optimized biofuel supply curve for the Western United States. *Biomass and Bioenergy* 34(11), 1597–607. doi:10.1016/j. biombioe.2010.06.007.

Pine, J. B. 1993. *Mass Customization: the New Frontier in Business Competition*. Harvard Business School Press, Boston, p. 333.

Pingoud, K., Pohjola, J. and Valsta, L. 2010. Assessing the integrated climatic impacts of forestry and wood products. *Silva Fennica* 44(1), 155–75. doi:10.14214/sf.166.

Porter, M. E. 2000. Location, competition, and economic development: local clusters in a global economy. *Economic Development Quarterly* 14(1), 15–34. doi:10.1177/089124240001400105.

Puettmann, M., Bergmann, R., Hubbard, S., Johnson, L., Lippke, B., Oneil, E. and Wagner, F. G. 2010. Cradle-to-gate life-cycle inventory of US wood products production: CORRIM Phase I and Phase II products. *Wood and Fiber Science* 42, 15–28.

Raunikar, R., Buongiorno, J., Turner, J. A. and Zhu, S. 2010. Global outlook for wood and forests with the bioenergy demand implied by scenarios of the Intergovernmental Panel on Climate Change. *Forest Policy and Economics* 12(1), 48–56. doi:10.1016/j. forpol.2009.09.013.

Published by Burleigh Dodds Science Publishing Limited, 2020.

Raymer, A. K., Gobakken, T., Solberg, B., Hoen, H. F. and Bergseng, E. 2009. A forest optimisation model including carbon flows: application to a forest in Norway. *Forest Ecology and Management* 258(5), 579–89. doi:10.1016/j.foreco.2009.04.036.

Riala, M. and Ilola, L. 2014. Multi-storey timber construction and bioeconomy – barriers and opportunities. *Scandinavian Journal of Forest Research* 29(4), 367–77. doi:10.1 080/02827581.2014.926980.

Sathre, R. and Gustavsson, L. 2006. Energy and carbon balances of wood cascade chains. *Resources, Conservation and Recycling* 47(4), 332–55. doi:10.1016/j. resconrec.2005.12.008.

Scarlat, N., Dallemand, J.-F., Monforti-Ferrario, F. and Nita, V. 2015. The role of biomass and bioenergy in a future bioeconomy: policies and facts. *Environmental Development* 15, 3–34. doi:10.1016/j.envdev.2015.03.006.

Scharin, H. and Wallström, J. 2018. The Swedish $CO_2$ tax – an overview. prepared by Anthesis Enveco AB under a contract from Borg & Co AB for the Institute of Energy Economics, Japan (IIEJ). Available at: http://www.enveco.se/wp-content/uploads/2018/03/Anthesis-Enveco-rapport-2018-3.-The-Swedish-CO2-tax-an-overview.pdf.

Schuler, A. and Buehlmann, U. 2003. Identifying future competitive business strategies for the U.S. residential wood furniture industry: benchmarking and paradigm shifts. USDA Forest Service Northeastern Research Station. General Technical Report NE-304, 17p.

Sikkema, R., Dallemand, J. F., Matos, C. T., van der Velde, M. and San-Miguel-Ayanz, J. 2017. How can the ambitious goals for the EU's future bioeconomy be supported by sustainable and efficient wood sourcing practices? *Scandinavian Journal of Forest Research* 32(7), 551–8. doi:10.1080/02827581.2016.1240228.

Sims, R. E. H. 2003. Bioenergy to mitigate for climate change and meet the needs of society, the economy and the environment. *Mitigation and Adaptation Strategies for Global Change* 8(4), 349–70. doi:10.1023/B:MITI.0000005614.51405.ce.

Skullestad, J. L., Bohne, R. A. and Lohne, J. 2016. High-rise timber buildings as a climate change mitigation measure – a comparative LCA of structural system alternatives. *Energy Procedia* 96, 112–23. doi:10.1016/j.egypro.2016.09.112.

Sohngen, B. and Sedjo, R. 2000. Potential carbon flux from timber harvests and management in the context of a global timber market. *Climatic Change* 44(1/2), 151–72. doi:10.1023/A:1005568825306.

Song, M. and Gazo, R. 2013. Competitiveness of US household and office furniture industry. *International Journal of Economics and Management Engineering* 3(2), 47.

Stump, B. and Badurdeen, F. 2012. Integrating lean and other strategies for mass customization manufacturing: a case study. *Journal of Intelligent Manufacturing* 23(1), 109–24. doi:10.1007/s10845-009-0289-3.

Toppinen, A., Röhr, A., Pätäri, S., Lähtinen, K. and Toivonen, R. 2018. The future of wooden multistory construction in the forest bioeconomy – a Delphi study from Finland and Sweden. *Journal of Forest Economics* 31, 3–10. doi:10.1016/j.jfe.2017.05.001.

Trømborg, E., Buongiorno, J. and Solberg, B. 2000. The global timber market: implications of changes in economic growth, timber supply, and technological trends. *Forest Policy and Economics* 1(1), 53–69. doi:10.1016/S1389-9341(00)00005-8.

Um, J. 2017. Improving supply chain flexibility and agility through variety management. *The International Journal of Logistics Management* 28(2), 464–87. doi:10.1108/IJLM-07-2015-0113.

Published by Burleigh Dodds Science Publishing Limited, 2020.

Van de Kuilen, J. W. G. V. D., Ceccotti, A., Xia, Z. and He, M. 2011. Very tall wooden buildings with cross laminated timber. *Procedia Engineering* 14, 1621-8. doi:10.1016/j. proeng.2011.07.204.

Wagner, E. R. and Hansen, E. N. 2005. Innovation in large versus small companies: insights from the small companies US wood products industry. *Management Decision* 43(6), 837-50. doi:10.1108/00251740510603592.

Wang, L., Toppinen, A. and Juslin, H. 2014. Use of wood in green building: a study of expert perspectives from the UK. *Journal of Cleaner Production* 65, 350-61. doi:10.1016/j.jclepro.2013.08.023.

Waugh, A., Wells, M. and Lindegar, M. 2010. Tall timber buildings: application of solid timber constructions in multi-storey buildings. *Proceedings of the International Convention of Society of Wood Science and Technology and United Nations Economic Commission for Europe - Timber Committee*, 11-14 October 2010, Geneva, Switzerland Paper, 9p.

Wear, D. N., Dixon, E., Abt, R. C. and Singh, N. 2015. Projecting potential adoption of genetically engineered freeze-tolerant Eucalyptus in the United States. *Forest Science* 61(3), 466-80. doi:10.5849/forsci.14-089.

Winistorfer, P. M. 2005. Competitiveness, manufacturing, and the role of education in the supply chain. *Forest Products Journal* 55(6), 6-16.

Winjum, J. K., Brown, S. and Schlamadinger, B. 1998. Forest harvests and wood products: sources and sinks of atmospheric carbon dioxide. *Forest Science* 44(2), 272-84.

Wolfshlehner, B., Linser, S., Pulzi, H., Bastrup-Birk, A., Camia, A. and Marchetti, M. 2016. Forest bioeconomy- a new scope for sustainability indicators. *European Forest Institute*. EFI Central-East and South-East European Regional Office (EFICEEC-EFISEE), University of Natural Resources and Life Sciences, Vienna, Austria, 31p.

# Chapter 2

## Agroforestry for hardwood timber production

*J. W. 'Jerry' Van Sambeek, formerly of USDA Forest Service Northern Research Station and University of Missouri Center for Agroforestry, USA*

## 1 Introduction

Managing hardwood trees for high-quality sawlogs within the agroforestry practices of alley cropping, silvopasture, forested riparian buffers and upland (windbreaks) buffers means intensively managing relatively wide-spaced trees and a ground cover as a companion crop on the same unit of land. These agroforestry practices provide the opportunity to produce logs and other tree products as a long-term investment while obtaining regular income through the production of companion crops or animals to offset some of the carrying costs associated with establishing the practices. The introduction of hardwoods into agricultural cropping systems also provides important ecosystem services and environmental benefits that conservation farming practices alone do not provide such as moderating air and soil temperatures and increasing biological diversity both above and belowground (Stamps et al., 2002; Jose, 2009; Kremer and Kussman, 2011). By moderating the microclimate, animals stay cooler and may graze longer in a silvopastural practice (Jose et al., 2017). Although forage yields may decrease in the shaded environment, the duration of forage production may be longer (Lehmkuhler et al., 1999; Kallenbach,

http://dx.doi.org/10.19103/AS.2018.0041.16

2009). In addition, forage quality will be improved, especially for crude protein content (Lin et al., 2001; Kallenbach et al., 2006; Pang et al., 2019b).

Alley cropping in the temperate regions will typically involve managing an annual grass or forage crop between rows of trees possibly with a perennial ground cover within the trees' rows. The relatively open tree spacing should allow adequate sunlight to reach the companion crops. Open-grown trees (ideally between 40% and 60% canopy coverage) are managed so that there is adequate light for the forage and vertical distribution of roots to minimize root competition for water and soil nutrients. Recent research indicates 50% of ambient sunlight for many crops and forages results in statistically similar yields as growing in full sun (Lin et al., 1999; Semchenko et al., 2012; Pang et al., 2019a). Differences in shade tolerance when measured under tree canopies are apparently less likely to be in response to reduced light and more likely a function of root competition for water and nutrients, especially available soil nitrogen (Jose et al., 1997). The addition of nitrogen-fixing shrubs within-tree rows may help by adding nitrogen to the soil and providing side shade to improve stem quality of the crop trees (Schlesinger and Williams, 1984).

Silvopasture deliberately combines forage and cattle with trees as part of a larger, managed intensive grazing system that includes open pasture (Kallenbach, 2009). Silvopasture designs should include open pastures as a place to move livestock while grazed forages are allowed to recover or when soils are high in available water to minimize soil compaction and damage to tree roots. Agroforestry practices that include rotational grazing or haying can reduce the rooting zone of the forages and expand the rooting zone for the deeper tree roots (Jose et al., 2017). Overgrazing by removing 50–80% of a plant's forage rapidly reduces root growth by 5–100%, respectively, and substantially increases the time (normally 3–4 weeks) needed for rest and regrowth (USDA-NRCS, 2005).

Buffers can also be managed for timber production. Forested riparian buffers managed with open-grown trees and a grass understory are frequently more effective in reducing surface run-off and capturing nutrients and sediments than buffers with dense tree plantings next to the waterway and an adjacent grass strip. Effective windbreaks with three or more rows of tall trees and an ideal canopy density of 40–60% will allow sufficient light penetration to maintain an herbaceous ground cover (Cunningham, 1988). Sheltered zones between two and twenty times tree heights can result in overall higher crop yields by reducing wind speeds and creating favourable microclimates especially under droughty conditions (Kort, 1988; Rivest and Vezina, 2015). Open-grown trees will require more human intervention to produce high-quality logs in buffers unlike in densely spaced tree plantings where side shade forces vertical growth on a straight central stem and early death and shedding of relatively small diameter branches.

## 2 Impact of ground cover on tree growth

The ground cover seeded as a crop or forage can substantially affect the growth of hardwood trees in an agroforestry practice with differences due to the type of cover, the competitiveness of the trees and soil or climatic conditions (Van Sambeek and Garrett, 2004). Usually the best tree growth is achieved with removal of all understory vegetation either by using chemical or mechanical methods or by mulching. Most agroforestry practices are installed on unlevelled ground, so vegetation-free approaches are not ecologically viable approaches. Establishment of some legumes as living mulches can be almost as effective as chemical or mechanical methods to reduce ground cover competition, especially with hardwoods that respond well to high-nitrogen fertilizers (Van Sambeek et al., 1986; Alley et al., 1999). In contrast, some perennial grasses established as living mulches can result in low foliage nitrogen concentrations in the trees and very slow growth (Schlesinger and Van Sambeek, 1986; Miller et al., 1987).

In a study to determine how competitive herbaceous legumes and grasses were, Van Sambeek and Garrett (2004) synthesized the results from over 110 research reports or papers that reported on hardwood tree growth with different ground covers. To be included in the database, these reports or papers had to include the growth rate for one or more tree species for stem diameter and/or height for 2 or more years after establishing a single species ground cover (Van Sambeek, 2010). Reports also had to include information from control plots with trees growing in vegetation-free areas or in areas with a living mulch of mixed vegetation including short-lived herbaceous legumes, non-legume broadleaved forbs and grasses. A living mulch of weeds reduces tree growth to approximately 60% of that reported for trees growing in adjacent vegetation-free plots with minor differences among the hardwoods grown for high-quality logs and wood products (Table 1). Insufficient data were found to compare effects of controlled grazing although it is likely similar to mowing of mixed vegetation types that further reduces tree growth an additional 16% for black walnut and an additional 10% for other hardwoods (Van Sambeek and Garrett, 2004).

Depending on the legume species and agroforestry practice, hardwood tree growth can either be impacted quite severely or slightly improved with a legume companion crop (Table 1). A living mulch of forage legumes has less of an impact on hardwood tree growth than does a living mulch of weeds (White et al., 1981; Van Sambeek and Garrett, 2004). Annual legumes such as soybeans (*Glycine max*) usually have less of an impact than perennial legumes with some exceptions like crownvetch (*Coronilla varia*), arrowleaf (*Trifolium vesiculosum*) and subterranean (*T. subterraneum*) clovers. Although available soil nitrogen is the nutrient most likely to be deficient, nitrogen-fixation rates of legumes

**Table 1** Average tree growth rate and standard deviation as percent of growth rates when grown with and without different ground covers and average-reported nitrogen-fixation rates for the different ground covers

| Legume or grass ground cover | | Growth reduction (%) | N-fixation rates (kg/ha) |
|---|---|---|---|
| Trifolium vesiculosum | Arrowleaf clover | 122 + 54 (10)[a] | ??? |
| Glycine max | Soybean | 103 + 18 (8) | 160 |
| Trifolium subterraneum | Subterranean clover | 94 + 54 (18) | 145 |
| Coronilla varia | Crownvetch | 93 + 43 (28) | 230 |
| Mixed genera and species | Annual grasses | 93 + 20 (12) | NA |
| Brassica napus | Kale | 89 + 20 (12) | NA |
| Muhlenbergia schriberi | Nibblewill | 88 + 25 (15) | NA |
| Trifolium incarnatum | Crimson clover | 81 + 44 (34) | 95 |
| Vicia sativa | Hairy vetch | 79 + 33 (172) | 135 |
| Agrostis giganta | Redtop | 78 + 18 (10) | NA |
| Festuca rubra | Red fescue | 73 + 26 (82) | NA |
| Poa pratensis | Bluegrass | 72 + 27 (27) | NA |
| Bromus inermis | Bromegrass | 72 + 14 (12) | NA |
| Trifolium pratense | Red clover | 71 + 28 (122) | 125 |
| Trifolium ambiguum | Kura clover | 71 + 13 (13) | ??? |
| Triticum and Hordeum spp. | Winter cereal grains | 68 + 39 (15) | NA |
| Lespedeza cuneata | Sericea lespedeza | 67 + 16 (33) | 175 |
| Trifolium repens | White clover | 67 + 19 (56) | 170 |
| Mixed genera and species | Mix of grasses and legumes | 67 + 24 (80) | ??? |
| Phleum pratense | Timothy | 66 + 13 (11) | NA |
| Dactylis glomerata | Orchard grass | 65 + 28 (103) | NA |
| Lotus corniculatus | Bird's foot trefoil | 65 + 23 (44) | 146 |
| Trifolium fragiferum | Strawberry clover | 65 + 33 (57) | ??? |
| Mixed genera and species | Mix of forbs without legumes | 64 + 19 (17) | NA |
| Kummerowia striata | Striate lespedeza | 63 + 14 (5) | ??? |
| Mixed genera and species | Mix grasses, forbs, legumes | 63 + 29 (195) | ??? |
| Mixed genera and species | Perennial grasses | 63 + 20 (20) | NA |
| Mixed genera and species | Mowed mixed vegetation | 61 + 25 (38) | ??? |
| Mixed genera and species | Mix of legumes only | 59 + 22 (28) | ??? |
| Lolium perenne | Ryegrass (annual/perennial) | 59 + 17 (10) | NA |
| Kummerowia stipulacea | Korean lespedeza | 59 + 12 (36) | ??? |
| Mixed genera and species | Native legumes | 55 + 24 (24) | 50 |
| Mixed genera and species | Ferns | 54 + 19 (11) | NA |
| Mixed genera and species | Mix of grasses only | 52 + 22 (14) | NA |
| Medicago sativa | Alfalfa | 51 + 24 (15) | 190 |
| Kummerowia striata | Common lespedeza | 45 + 20 (10) | ??? |
| Schedonorus arundinaceus | Tall fescue | 43 + 28 (141) | NA |

[a] (#) = number of comparisons found in the literature, field reports and unpublished datasets.
NA – not applicable; ??? – insufficient information to provide an estimated rate of atmospheric nitrogen fixation for these legumes.

are poorly correlated with average reduction in tree growth (Table 1). Few studies have measured response of walnut or other hardwoods to harvesting of legumes for forage, although it can be assumed much of the fixed nitrogen is removed with the forage resulting in no net gain in available soil nitrogen to the trees (Van Sambeek, 2017).

In general, grass sods are more competitive than forage legumes as living mulches, although there is considerable variation among the different species of grasses and some variation among hardwood species (Van Sambeek and Garrett, 2004). When compared to tree growth in vegetation-free plots, reductions in tree growth with grasses average 47% for black walnut compared to only 32% for other hardwoods. In general, the least competitive grasses including several turf grasses and the annual cereal grains reducing hardwood growth as little as 30% compared to growth for trees in vegetation-free plantings. It is not uncommon to find reports of hardwood tree growth being reduced by more than 70% of the potential growth with a companion crop of tall fescue (*Schedonorus arundinaceus*) (Todhunter and Beineke, 1979; Dey et al., 1987; Van Sambeek et al., 1989; Van Sambeek and McBride, 1991; Alley et al., 1999). Reducing hardwood tree growth to 30% of their potential without competing vegetation has the effect of increasing rotation length by threefold, that is instead of 50-60 years (assuming 1 cm dbh/year) for harvestable veneer logs, rotation lengths will likely exceed to 170–200 years.

It is unclear why tall fescue, a relatively deep-rooted, sod-forming grass used in many silvopastoral practices, is so competitive, although there is evidence the dense fibrous root system is quite competitive for available soil nitrogen essentially starving the trees for nitrogen (Van Sambeek et al., 1989). There is good evidence that phytotoxic compounds may also be released from tall fescue roots or decomposing litter that slow growth of black walnut and white ash seedlings (Rink and Van Sambeek, 1985, 1987; Preece et al., 1991). Reduced growth of hardwoods by phytotoxic compounds for other grasses and forbs is also relatively well documented (Larsen and Schwarz, 1980; Ponder, 1986; Adler and Chase, 2007; Van Sambeek, 2017). Likewise, reduced growth of companion crops by trees producing phytotoxins is also well documented (Rietveld, 1982, 1983; De Scisciolo et al., 1990; Jose and Gillespie, 1998; Rizvi et al., 1999; Willis, 2000; von Kiparski et al., 2007).

## 3 Growing-space requirements

Unless thinned, hardwood stands will tend towards a condition called fully stocked meaning the site is being fully utilized. Stocking (or density) is a measure of the number of trees per ha compared with an optimal density for balanced health and growth (Stewart and Dawson, 2013). Stocking guides for upland temperate hardwoods when displayed graphically indicate if the trees are fully utilizing a site as overstocked, fully stocked or understocked (Gingrich, 1967). Unfortunately stocking charts are not very useful for evaluating tree-to-tree competition in most agroforestry practices because they would be considered understocked if an herbaceous understory is to be maintained. In addition, stocking guides were developed for stands where height growth is forced by side shade and trees have relatively small crowns compared to open-grown trees due to early death of shaded small diameter limbs in the lower crown.

Tree spacing within an agroforestry practice will likely be a compromise between having open-grown trees with minimal shading of the companion crop and denser spacing providing side shade to force tree height growth on a single stem and early death of small diameter branches. Krajicek et al. (1961), for planted trees, proposed that the ratio of the sum of the theoretical crown areas of all trees in a defined area when dividing by that area would better quantify tree competition than would stocking charts. Krajicek (1966) also found there was a strong linear relationship between stem diameter at breast height (DBH) and the crown width (CW) on open-grown temperate hardwoods (Table 2). For example, the CW in metres for open-grown black walnut trees is:

$$CW \text{ (in m)} = (0.239)(DBH \text{ in cm}) + 1.49$$

as derived from the English version of this equation:

$$CW \text{ (in feet)} = (1.99)(DBH \text{ in inches}) + 4.87$$

The crown competition factor (CCF) was coined by Krajicek et al. (1961) to describe the ratio between the sums of the open-grown crown area of all trees divided by the area the trees occupy. Because original calculations were made with crown area as the per cent of the area, the ratio is now multiplied by 100 to give per cent of area covered by tree crowns. The theoretical crown area rather than the actual crown area is used because actual crown areas when canopies

**Table 2** Equations for determination of crown widths in metres as a function of diameter at breast height (DBH, 1.4 m tree height) in centimetres for several open-grown temperate agroforestry timber species (converted from English units in listed sources)

| Species | Crown width (m), DBH (cm) | $R^2$ | Sources |
|---|---|---|---|
| Black walnut | 0.24 DBH + 1.49 | 0.99 | Krajicek (1967), Krajicek (1966) |
| White oak | 0.23 DBH + 0.55 | 0.99 | Krajicek et al. (1961) |
| Red/black oaks | 0.20 DBH + 1.38 | 0.98 | Krajicek et al. (1961) |
| Pin oak | 0.18 DBH + 2.76 | 0.91 | Krajicek (1967) |
| Mixed oak/hickory | 0.22 DBH + 0.95 | 0.98 | Krajicek et al. (1961) |
| Sweetgum | 0.24 DBH + 0.81 | 0.93 | Krajicek (1967) |
| Bald cypress | 0.24 DBH + 2.53 | 0.89 | Krajicek (1967) |

close adjust to the growing space and are no longer linearly related to stem diameter. For example, the CCF for a 2-ha silvopasture paddock with 50 black walnut trees with an average DBH of 40 cm is calculated as:

$$CCF = \frac{(100)(3.14)(\# \text{ trees})(0.12\, DBH + 0.75)^2}{(10000)(\#ha)}$$

where # is the number of trees and ha is the planting area. The resulting CCF is 25 indicating 25% of the paddock is covered by the sum of the theoretical crown areas for the 50 trees or the ratio of 0.25 $m^2$ of tree crowns per $m^2$ of pasture.

Schlesinger (1997) compared the CCF and ratio of stem diameter growth of walnut in close-spaced plantings to diameter growth in adjacent wide-spaced planting to estimate when tree competition began. Regression analyses indicated diameter growth started to be affected at a CCF of approximately 80, well under the original assumptions made by Krajicek et al. (1961) that between-tree competition begins when canopies close. Because root spread can be two to three times greater than canopy spread, most likely tree root competition for soil water and nutrients is more important than canopy competition for light in determining effects of tree competition (Schlesinger, 1988b).

Management recommendations as to when to thin can be made with reference to an appropriate range of CCF values. For example, maintaining black walnut stocking between CCFs of 100 and 120 is recommended for veneer production to keep trees growing near maximum rates while having some competition to promote good stem form and the slow growth and early death of lower limbs (USDA-FS, 1981). For nut production, CCFs of between 80 and 100 are recommended to maintain open canopies with good light penetration to the side branches (Jones et al., 1998; Reid et al., 2009).

Schlesinger (unpublished) found a CCF = 285 is equivalent to 100% stocking (net annual grow is zero) in pure walnut stands and a CCF of 100 is equivalent to 35% stocking. At 35% stocking, Blizzard et al. (2013) found transmitted light averaged approximately 40% of full sunlight. This is above light levels where shading alone decreases forage production (Pang et al., 2019a). Gardner et al. (1985) reported that photosynthesis equals or exceeds daily respiration for most forages at about 10% of full sun with light saturation for cool- and warm-season forages at 50% and 85% of full sun. From the above relationships, charts have been produced to graphically determine CCF (Fig. 1) or crown cover using the stocking chart format developed by Gingrich (1967) to relate tree density, tree DBH and basal area to CCF or tree crown cover as a per cent of the total area (USDA-FS, 1981; Law et al., 1994).

In some agroforestry practices, tree spacing can have a greater distance between rows than within rows. In this situation, tree competition affecting

(a)

(b)

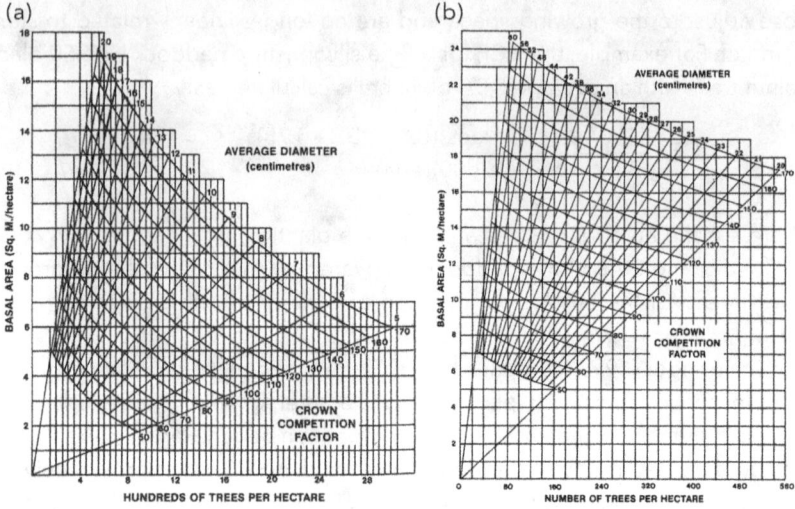

**Figure 1** Charts for graphical determination of CCF or crown competition factor (USDA-FS, 1981). (a) Average tree diameters 5-20 cm. (b) Average tree diameters 20-60 cm.

stem growth is assumed to begin when the crowns begin to touch. When tree crowns touch, the distance between trees in metres is approximately 2.4 m plus 0.11 times the average of their DBHs in cm. For example, two trees 10 and 20 cm DBH if more than 4.1 m apart should still be putting on maximum diameter growth. The relationship can also be used to determine the distance between trees when thinning to allow trees to sustain maximum growth until the next planned thinning. For example, if trees are annually growing 1 cm in diameter and the next thinning is anticipated in 10 years, the above trees would need to be thinned to 5.2 m apart to sustain current growth rates to the next thinning.

In most agroforestry practices, trees will be thinned before canopy closure and substantial between-tree competition occurs to maintain adequate light to the understory crops or forages. Before thinning, the CCF should be calculated to determine how intense between-tree competition is. When managing hardwoods for high value lumber and timber products, CCF should be maintained between 90 and 120. When CCFs are near 120 or higher, removing one-third of the trees by selective thinning of the slower growing, poorly formed trees is usually sufficient to allow remaining trees to grow near maximum rates for another 5-10 years. The higher the CCF above 120 when thinned, the longer it may take for trees to respond with increased growth rates. Because of wide genetic variability in growth rates among individual trees of the same species, row thinning is not recommended for hardwoods unless trees have been grafted with superior timber or timber/nut scions. A double-row alley-cropping system has been suggested for managing grafted pecan and walnut during establishment that has narrow perennial grass alleyways allowing season-long access to the trees altering with wide alleyways for crops or forages (Van Sambeek and Reid, 2017). The double-row design allows row thinning on the diagonals to maintain open-crowns for nut production and light penetration into the cropped alleyways.

An alternative method to thinning hardwoods in a silvopastural or multirow buffer practice is crop tree management. This system focusses on managing the growing space around the canopy by releasing individual trees using a crown touching technique (Perkey et al., 1994; Miller et al., 2007). Theoretically up to 70–100 of the faster-growing trees with the potential for a high-quality log that are approximately 10 m or more apart are selected per ha and marked for release. Release is done simply by looking up into the crown of each crop tree and removing any adjacent trees that are touching the crop tree canopy, or that are likely to touch the crown of the crop tree before the next release. Crowns of temperate hardwoods typically expand 0.3 m/year. If the next release will be in 10 years, then the crown of the crop tree should be given 3 m or more distance from the crowns of the nearest competitive trees to permit crown expansion on three or four sides. If two crop trees are very close together, they can be treated as 'one' tree so both trees are released on three sides. Trees outside the release area around crop trees are left to compete with each other in part to keep crop trees from leaning to capture sunlight and to provide side shade to minimize production of epicormics sprouts on the crop trees.

## 4 Pruning recommendations and practices

In natural stands of close-spaced trees, side shade limits growth and longevity of lateral branches and slows stem diameter growth that allows time for dead branches to be shed and wounds to callus limiting defects to the central core. In any agroforestry practices with open-canopy trees, tree growth rates will be higher, branches larger in diameter and wound callus is likely to include dead stubs if branches are not artificially pruned. Pruning will be essential for production of high-quality logs of most species in any agroforestry practice. The objective should be clear, knot-free logs that peel easily for veneer that produce veneer sheets easily bookend-matched and have an attractive appearance due to straight grain and uniform cathedral patterns (Kesner, 1986). In production of high-grade construction lumber, knot-free timber is stronger with more uniform strength properties along with being easier to machine and finish.

To determine if artificial pruning is necessary or can be economically justified, Schlesinger (1986) analysed data on hardwoods in terms of their ability to shed branches naturally, how likely they were to produce epicormics sprouts in response to pruning and whether the average price difference between log grades (Prime $/mbf – Number 1 $/mbf and Number 1 $/mbf – Number 2 $/ mbf) justified pruning. Schlesinger's tabulated results are given in Table 3 for Appalachian hardwoods with price differentials adjusted to present values. Additional hardwood species important to the Midwestern United States were added based on the research by Stubbs (1986), Meadows (1995), Miller et al. (2007) and the author's own experiences.

Tree species with persistent live or dead branches must be artificially pruned to produce knot-free timber (Table 3, Shedding, Category D and E). Artificial pruning must be considered an investment to improve log quality so that it will command a price high enough to more than cover the compounded costs for pruning that occur early in the stand rotation. For most hardwoods, the price differential between log grades is not sufficient to justify artificial pruning for log quality only (Table 3). For hardwoods where logs graded as prime can be sold as veneer logs, the price differential may justify pruning, especially in an agroforestry practice where trees should be regularly pruned to increase light penetration to the understory vegetation. Kesner (1986) reported hardwood logs sold for veneer sell for three to five times more than if sold as high-quality sawlogs. Because of the price differential between veneer and sawlogs and the increased costs of pruning above 3 m, pruning operations should concentrate on producing 2.5–3 m long logs for veneer rather than 5.5–6 m for sawlogs. Other advantages of pruning for one high-quality veneer log on open-grown trees include being less susceptible to wind damage and more easily managed for nut production.

Wide variation exists among the hardwood species in the production of epicormics branches following pruning (Table 3). The probability that a tree will produce epicormics branches after pruning is a function of its position in the stand canopy, tree density before thinning, the clear bole to crown ratios and the amount of canopy removed. The probability of epicormic sprouting increases substantially when clear bole to crown ratios exceed a 40:60 ratio or if more than a quarter of the potential canopy is removed in a single pruning operation. Hardwood species listed in Table 3 that have a high probability of producing epicormics branches may require less intense, more frequent pruning when grown in an agroforestry practice to minimize bole sprouting.

How and when to prune hardwood trees is not without controversy. Current recommendations are to first concentrate on any necessary pruning to remove forks to maintain a central, straight stem to the height of a harvestable log(s), and then begin removing the largest diameter lateral branches (Schlesinger and Shigo, 1989). In dense plantings few trees may need to be pruned as poorly formed trees will be removed during an early thinning. In most agroforestry practices, however, fewer trees are planted and most hardwood trees may need to be repeatedly pruned to produce a knot-free central stem and maximize light penetration to the understory plants.

**Table 3** Hardwood species' characteristics relative to their ability to shed branches naturally, produce epicormics sprouts following pruning, log-grade price differential and whether cost-effective to prune

| Genus species | Common name | Shedding[a] | Sprouting[b] | Price[c] | Prune |
|---|---|---|---|---|---|
| Acer rubrum | Red maple | B | B/C | C | No |
| Acer saccharinum | Silver maple | - | B | B | No |
| Acer saccharum | Sugar maple | D | B/C | C | - |
| Betula alleghaniensis | Yellow birch | B | C | - | - |
| Betula lenta | Sweet birch | B | C | B | No |
| Carya glabra | Pignut hickory | B | C | B | No |
| Carya illinoensis | Pecan | B | B/C | B | No |
| Carya laciniosa | Shellbark hickory | B | C | B | No |
| Carya ovata | Shagbark hickory | B | C | B | No |
| Carya tomentosa | Mockernut hickory | B | C | B | No |
| Celtis occidentalis | Hackberry | - | - | - | No |
| Fagus grandifolia | Beech | B | A | - | No |
| Fraxinus americana | White ash | B | D | C | - |
| Fraxinus pennsylvanica | Green ash | B | D | C | No |
| Juglans cinerea | Butternut | C | - | E | Yes |
| Juglans nigra | Black walnut | E | B | F | Yes |
| Liriodendron tulipifera | Yellow poplar | B | C | B | No |
| Liquidambar styraciflua | Sweetgum | - | B | - | - |
| Magnolia acuminata | Cucumbertree | B | - | - | No |
| Nyssa sylvatica | Black gum | C | - | - | No |
| Platanus occidentalis | Sycamore | B | B/C | - | No |
| Populus spp. | Poplar/aspen | A | C | - | No |
| Prunus serotine | Black cherry | C | B | D | Yes |
| Quercus alba | White oak | C | A | E | Yes |
| Quercus bicolor | Swamp white oak | D | B | D | - |
| Quercus coccinea | Scarlet oak | D | - | D | Yes |
| Quercus falcata | Cherrybark oak | D | A/B | E | Yes |
| Quercus macrocarpa | Bur oak | C | - | D | - |
| Quercus michauxii | Chestnut oak | C | B | D | No |
| Quercus rubra | Northern red oak | B | A/B | D | - |
| Quercus shumardii | Shumard oak | C | B | D | No |
| Quercus velutina | Black oak | C | - | D | Yes |
| Robinia pseudoacacia | Black locust | B | - | - | No |
| Sassafras albidum | Sassafras | B | - | - | No |
| Tilia americana | Basswood | B | B | A | No |
| Ulmus Americana | American elm | B | A | - | No |

[a] Natural shedding categories: A = excellent, B = good under side shade, C = okay, D = poor and E = very poor.
[b] Categories for producing epicormics sprouts: A = very high, B = high, C = moderate and D = low.
[c] Price differential lumber/MBF were A = US$0-50, B = US$50-100, C = US$100-150, D = US$150-200, E = US$200-250 and F = >US$250.
Source: modified from Schlesinger (1986).

A fork usually consists of two or more closely spaced, vertically oriented stems with diameters that are within half the diameter of the largest stem. If forking occurs in the current year's growth (sometimes called a crow's nest), artificial pruning may not be required (Schlesinger and Bey, 1978). Schlesinger (1982) evaluated effectiveness of pruning a crow's nest to a single terminal shoot in black walnut saplings and found no differences in percentage of trees that eventually developed straight, single stems whether pruned or left to naturally straighten. Most of the competing terminal shoots in a crow's nest became lateral branches in the next growing season. Saplings with vertically oriented stems having two or more years of older tissues are less likely to naturally straighten and will benefit from corrective pruning to a single terminal shoot (Beineke, 1977). Unwanted vertical stems should be cut diagonally between the unwanted stem and the branch bark ridge (described below) and exiting at a depth equal to the base of the branch bark ridge (Shigo, 1989).

In some cases, especially when open grown, saplings develop multiple forks and crooks in response to disease, injury or poor genetics. For these saplings, coppicing is frequently recommended; however, there is no consensus whether to coppice the stem close to the root collar or to high coppice just below the deformed stem. Because coppiced saplings have well-developed root systems, coppice growth can be very rapid (and may require staking) with little net loss in height growth. Initially new shoots of coppiced saplings will have excellent form but most saplings will again develop form issues after the first or second growing season. Results are also mixed whether the coppice cut should be a horizontal cut or a sloping cut made just above a dormant bud. A sloping high-coppice cut behind a bud tends to produce the fewest sprouts especially if made in the region of stem between where height growth ended each year. On pole-sized trees that have been cut horizontally for bark grafts, the stem naturally abscises a triangular piece of stem immediately behind a successful graft suggesting a sloping cut should be used for a high coppice.

Pruning should concentrate secondly on removing the fastest-growing or largest diameter lower lateral branches before removing the lowest lateral branches to raise the base of the crown (Shigo, 1989). For veneer-quality logs, it is best if lateral branches can be removed before they are 3–5 cm in diameter. Branches tend to cluster in the areas immediately below where a terminal bud was set at the end of each growing season. Multiple large diameter lateral branches in the same area should be slowly removed over multiple years to reduce the possibility of developing ring shake in the future log. If possible also avoid pruning large lateral branches in vertical alignment within 1 m of each other. Also removing more than a quarter of the next growing seasons' anticipated foliage is likely to substantially slow tree growth and induce epicormic sprouts to replace the lost crown (Schlesinger, 1988a). If more large diameter branches exist than can be removed in one pruning operation without harming the tree, diameter growth can be slowed by cutting or heading back branches to approximately a third of their original length. Numerous dormant and lateral buds will elongate the following spring feeding the entire tree with minimal increase in branch diameter. Also, if branch and stem are similar in diameter, head back the branch and wait a few years before removing stubbed branches to minimize main stem breakage.

When pruning lateral branches, there is the temptation to 'flush cut' or cut vertically close to the stem to make the stem smooth and begin laying down straight-grained wood sooner. Shigo et al. (1979) found flush-cut pruning tends to lead to substantial internal decay and other defects. The alternative is 'target' or 'collar' pruning. In his article on how branches are attached to stems, Shigo (1985) reported that in the spring the branch produces new xylem first which extends back along parts of the stem before the stem produces new xylem that extends into the branch (Fig. 2). Where the two tissues overlap is a slightly swollen area or collar around the branch. Within the branch collar are cells pre-programmed to rapidly form a compartmentalization zone that naturally blocks entrance of most microorganisms and insects into the stem when branches either die or are pruned (Shigo, 1984).

Proper target pruning just outside the branch bark ridge and branch collar (line AB in Fig. 3) results in a full circle of visible callus along the edges of the wound that will eventually cover the wound. O'Hara (2007) argues pruning

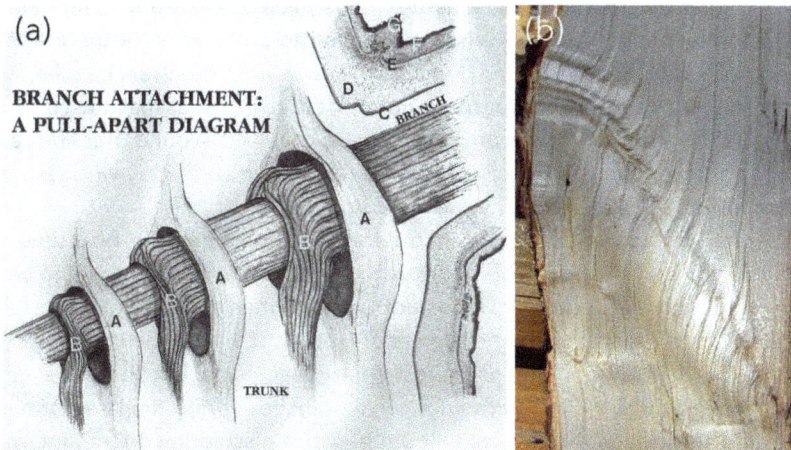

**Figure 2** (a) Exploded diagram showing how branches are attached to stems when new branch wood or collar (B) is overlaid by new stem wood or collar (A) resulting in a swollen branch collar laid down by the cambium (C) and the branch bark ridge (G) (Shigo, 1996) that forms as branch and stem tissues meet above the branch. D, E and F are the phloem, bark cambium (phellogen) and dead bark tissues, respectively. (b) Maple stem split through the region near a branch showing the overlapping growth rings.

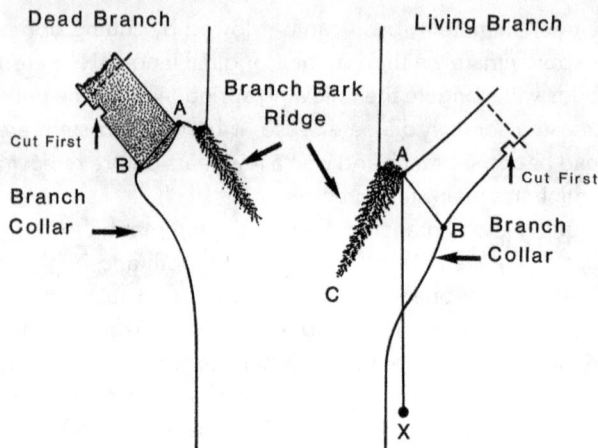

**Figure 3** Diagram of natural target pruning where pruning cuts are made at A-B just outside the branch collar on dead branches and along A-B on live branches where A-X is a vertical line just outside the branch bark ridge and the angle XAC is equal to XAB.

just outside branch bark ridge with vertical cuts through the branch collar (line from A emerging between X and B in Fig. 3) may produce a larger wound, but it will callus over faster and begin laying down clear wood sooner than target pruning especially on small diameter branches. Flush cutting (line AX in Fig. 3 if placed before the branch bark ridge) creates an even larger wound and removes the natural defence tissues found in the branch collar and delays callus formation leaving wood tissues exposed to pathogens for the longest period of time. Stub cutting, or pruning well away from the branch collar, will delay wound closure and increase the number of growth rings that must be laid down before straight wood is produced over the wound. Wound dressings are no longer recommended because they are expensive and rarely work as well as the tree's natural defences (Shigo and Shortle, 1984).

Although there is little published information on when the best time to prune is, the consensus is that hardwoods can be pruned anytime of the year except possibly the period between budburst and when the first leaves are fully mature (April-May when energy reserves are lowest). Dead branch pruning can be done anytime if living tissues in the branch collar are not cut creating a fresh wound (Schlesinger, 1988a). Pruning of live branches during the late dormant season usually leads to the least amount of wood discolouration and internal decay and the most rapid wound closure (Smith, 1980; Armstrong et al., 1981). Trees of some species should not be pruned when insects are present that are attracted to open wounds (Bedker et al., 2012). Although it should be avoided, there is little evidence that pruning trees that bleed after sap starts rising in the spring will result in additional discolouration or slow tree growth.

# 5 Log and wood quality

Straight, cylindrical butt logs consisting of perfect wood with straight grain, evenly spaced growth rings ought to be the objective for hardwood trees managed in an agroforestry planting. Few studies have apparently been done to see if we can achieve this objective. Based on what little information was published before 2003, Garrett (2003) suggested that the quality of hardwoods grown under alley cropping if properly managed would not vary substantially from forest-grown trees but the time required to grow trees to market size in an agroforestry practice would be greatly reduced. Theoretically, research comparing wood quality of planted hardwoods maintained at relatively low CCFs to trees harvested from natural stands should provide a clue as to what differences, if any, are likely to occur when open-grown hardwoods are managed for quality in an agroforestry practice. As is the case with many of these studies, Phelps and Chen (1989) found the open-grown planted hardwoods exhibited more defects mainly due to lack of intensive management.

Because of the price differential between veneer and sawlogs for most hardwoods, Kesner (1986) suggested managing trees for veneer should be more profitable even if the quality must be higher than managing for sawlogs. He proposed that logs marketed for veneer must meet six general requirements that include having a nearly cylindrical shape with a well-centred heart, a minimum of internal defects which are concentrated in either the central core or slab zone, uniform wood colour and texture and properly sawed to length without embedded soil and other debris. The last requirement is controlled by the logger at the time of harvest and will not be addressed in this section.

In an agroforestry practice with open-grown trees, the shape of the butt log is largely a function of regular pruning to maintain a central leader by removing forks. Cutter (1991) reported only minor differences in stem eccentricity of black walnut logs harvested from an alley-cropping practice where trees with elliptical crowns were grown in close spacings within rows and wide spacings between rows. Because eccentricity was measured as the radius between pith and outer growth ring, it can be assumed these trees also had a well-centred heart. With frequent and regular pruning, branch wounds should callus over quickly laying down straight-grained wood, thus limiting defects associated with branches that extend to the central core. Crook and sweep, log defects associated with forks, tend to disappear over time due to differential growth along the stem, and may potentially yield a round cylindrical stem with irregular grain limited to the central core along with the wavy pith centre. In his review on stem form, Larson (1963) concluded open-grown trees are unfortunately likely to have greater clear bole taper which will affect the grain appearance in lumber and veneer. He also concluded that open-grown trees in response to wind may exhibit more butt flare, so veneer-quality butt logs may need to be bucked a little longer if only straight-grained veneer is desired.

Carpenter et al. (1989) described a wide range of defects that can occur in hardwood logs including limbs, bumps and overgrowths; suppressed buds and epicormic sprouts; shakes and splits; holes including birdpeck; and decay and discolourations. Most of the important blemishes in wood are associated with abnormal features visible in the bark. As an aid to growers and timber buyers, Rast co-authored several photographic guides that included stereo pairs of photographs of external indicators of the above internal defects for northern red oak (Rast, 1982), black cherry (Rast and Beaton, 1985), black walnut (Rast et al., 1988), white oak (Rast et al., 1989), yellow poplar (Rast et al., 1990a), sugar maple (Rast et al., 1990b) and yellow birch (Rast et al., 1990c).

The most frequent bark distortions are bumps and overgrowths (catfaces) covering knots from lateral branches. Bumps are protuberances usually projecting over sound or rotting limb stubs while overgrowths appear as roughly concentric rings in the bark formed by the callus that has grown over knots in the wood. As the knot is buried deeper into the stem, overgrowths gradually disappear, especially knots from small diameter limbs. More bumps and overgrowths are reported on logs from open-grown trees than trees from natural stands (Phelps and Chen, 1989). On open-grown trees, lateral branches will persist for a longer time and have greater diameter growth without artificial pruning resulting in a larger area to callus over producing more prominent overgrowths that can remain visible in the bark when harvested even though straight-grained wood has been laid down over the wound callus.

Suppressed buds and epicormic branches can also result in bumps and overgrowths. Most suppressed buds are formed in the cambium at the base of each leaf at the time of shoot development (Kormanik and Brown, 1967). Although most axillary buds die, a few buds remain dormant within the meristem that can be observed as raised pimples on the wood surface when the bark is peeled off. Most of these dormant buds have a vascular trace that extends back to the primary xylem surrounding the pith. In cross section, the vascular trace is considered a defect in high-end veneers. Dormant buds or pinknots can also develop adventitiously following injury to the cambium. Overpruning, damage to the crown or other stresses can cause these suppressed buds to divide forming pinknot clusters, or the buds to elongate and develop into epicormics sprouts. Chen et al. (1995) found walnut lumber cut from an alley-cropping practice that was frequently pruned averaged four pinknots and four clusters of two to four pinknots per metre with minor differences between the basal and apical ends of the future veneer logs. Pinknots within clusters were usually within 3 cm of each other and usually associated with knots from pruned branches. Minor variation existed from lumber cut from trees of seed selected for nut production or timber production suggesting death of suppressed buds is controlled by environment and not genetics. Epicormic sprouts from dormant buds will leave small tapered knots in the stem until the sprout is shed or pruned off and the wound calluses over.

Ring and radial shakes are major defects because they cause wood to separate during processing for veneer or lumber. Both are associated with compartmentalization of wounds (Shigo, 1984). Wood in ring shakes separate along the circumferential plane when wood in a new growth ring becomes infected and separates from the underlying wood as the tree forms barrier zones to compartmentalize the infection. Shigo et al. (1979) reported that ring shake was associated with 11 of 17 pruning wounds following 'flush-cut' pruning of both live and dead branches. Ring shake is frequently found following removal of the branch collars that form around the bases of dying and dead branches on older trees.

Radial shakes are also often associated with wounds resulting from infected knots, insect attacks or mechanical injury. Some decay commonly extends upwards and downwards from the wound into the wood-producing longitudinal radial separations in the wood. When the wood is stressed such as during a drought or sudden changes in temperature, these separations are weak areas and will open resulting in vertical cracks in the bark followed by wound callus to close the crack (Butin and Shigo, 1981). Wound callus from repeated cracking when frozen wood is rapidly thawed by direct sunlight results in vertical bulges referred to as frost cracks. In an agroforestry practice with open-grown trees, stems of some species may need to be painted white to reflect sunlight during the winter when stems are not shaded by leaves. Trees with seams will not produce veneer-quality logs and should be removed when thinning. When young, trees with vertical seams should be coppiced below the seam hoping the new wounds will callus over without leaving weak areas that will become new radial shakes.

Holes are wounds that expose the sapwood and remain open or become occluded with wound callus leaving a depression in the bark (Carpenter et al., 1989). When logs are rotary-sliced, wounds do not show concentric growth rings in the wood as do overgrowths from knots. Common causes of wounds include mechanical damage from logging activities, increment borers, tap holes, boring insects and birdpeck. Trees in an agroforestry practice have an increased risk for developing holes associated with use of equipment or livestock to manage the understory crops and forages. New holes are usually accompanied by stain in the wood of the exposed growth ring which may separate leading to ring shake (Shigo, 1963, 1996). Sapsuckers are a common cause of holes as they drill into the sapwood to feed on bleeding sap and the sapwood frequently creating bands or rows of holes around the stem. Sapsuckers sample the cambium and sapwood of many trees and usually select particular trees to repeatedly tap for multiple years. Garrett et al. (1994) recommended that trees heavily damaged in an agroforestry practice by sapsuckers be retained when thinning to minimize risk to sapsuckers feeding on neighbouring trees after thinning.

Decay and discolouration in the wood are also defects that can be minimized through stand management. Smith (1980) found spring pruning wounds on walnut resulted in much smaller columns of discoloured wood and much more diffuse dark margins than did fall wounds. The dark discolouration is maintained in the heartwood and is a serious source of degrade in finished veneer. Eisner et al. (2002) suggested the amount of discoloured wood that develops following branch removal decreases as branch orientation changes from vertical to horizontal. The major cause of decay in stems and branches is infection of wounds by microorganisms that grow through the protective chemical barriers formed in the wood around wounds (Shigo, 1984; Hart, 1982).

Hoffard et al. (1977) and Hart et al. (1986) examined the economic impact and causes of decay in black walnut in the central United States and found presence of decay decreased log value by 66%. When stands were examined for probable cause of decay, over 40% had some damage caused by livestock, especially swine because swine scrape and peel the bark at the base of trees allowing decay to enter and develop. Kesner (1986) also indicated prolonged presence of livestock around trees usually results in dark stains in the wood and many small scars around the lower portion of the stem. Proper silvopastoral management using rotational grazing should minimize discolouration and decay in the wood associated with livestock, but not swine. Lehmkuhler et al. (2003) found no damage to four hardwood species by livestock when trees were planted into open pastures and protected by electric fencing used to create multiple paddocks for rotational grazing.

Few studies have been done comparing the appearance (texture) and properties of wood from forest-grown trees to open-grown trees in plantations or in an agroforestry practice (Cutter et al., 1997). Kesner (1986) and Cassens (2004) suggested the most attractive hardwood veneer is sliced from logs with six to ten rings per inch. Within agroforestry practices where tree CCF is maintained below 120 to minimize fluctuations in growth ring widths, texture is likely to be reduced. Presumably the price reduction for too few rings per inch will be offset by the increased volume of clear wood in trees from managed agroforestry practices. One option to improve texture is to plant seed from or graft saplings to selections from trees displaying unique figure such as 'bird's eye', 'fiddle-back' or 'curly grain' (Beals and Davis, 1977; Bragg and Stokke, 1999). Unfortunately, McKenna et al. (2015) found curly grain figure does not propagate true-to-type in grafts on open-grown walnut or within seedlings from parent trees with figure.

Agroforestry practices are unlikely to produce substantial changes in the colour of hardwood lumber or veneer. Nelson et al. (1969) found soil properties affected colour more than growth rate for fast- and slow-grown black walnut trees on good and poor sites in Indiana and Missouri. Later, Phelps (1989) reported that walnut heartwood from faster-grown trees is usually somewhat lighter in colour (luminance) than that from slow-grown trees. There is some indication that in heartwood an increase in specific gravity is likely to cause a darker colour.

The best indicators of easily measure parameters for accessing changes in important wood properties are probably wood-specific gravity or density (Phelps and Chen, 1999). In his review on stem form, Larson (1963) concluded that wood density is lower in the butt log of open-grown trees than in forest-grown trees. Cutter and Garrett (1993), however, found that height, diameter and wood-specific gravity was equal or higher for alley-cropped trees than forest-grown trees. Cutter (1991) found that specific gravity in the heartwood of 15-year-old black walnuts grown in an alley-cropping practice gradually increased from the pith towards the sapwood followed by a substantial decrease in specific gravity in the sapwood. Similarly, Phelps and Chen (1989) reported that specific gravity gradually increased in the heartwood with distance from the pith with a marked decrease in the sapwood although volumetric shrinkage of lumber cut from open-grown and forest-grown walnut was similar.

Although few studies have been done quantifying log and wood quality from open-grown hardwoods grown in an agroforestry practice, results comparing open-grown trees to forest-grown (slow-growth) trees would indicate that logs from an agroforestry practice are likely to have similar wood properties and colour, a less desirable texture (figure) and show more defects. Most defects could be limited to the central core if trees were properly pruned on a regular basis. At this point in time there is no reason not to accept Garrett's original conclusion in 2003 that the quality of hardwoods grown under an agroforestry practice (if properly managed) would not vary substantially from forest-grown trees but the time required to grow them to market size would be greatly reduced.

## 6 Summary

This chapter synthesizes available information on managing hardwood trees for production of veneer and high-quality sawlogs. Managing timber within the agroforestry practices of alley cropping, forested riparian and upland (windbreaks) buffers, and silvopasture typically involves planting hardwood trees at lower densities, and their development in more open conditions than which occurs in natural stands. Choice of crops or forages can substantially impact the growth rate of trees in the different agroforestry practices. Trees will need intensive management with frequent pruning and thinning to produce high-quality hardwood logs. Frequent thinning will be required to maintain open tree canopies to provide sufficient sunlight for crop and forage production. Frequent thinning will also be required to minimize within-tree species competition to promote production of wood that has evenly spaced growth rings and uniform texture. Target pruning should be done in the late dormant season to minimize wood discolouration and to create small wounds that callus over quickly, thus reducing the risk of infection and subsequent internal decay. Overall, high-quality open-grown hardwood trees can be produced in an agroforestry practice with intensive management, and in a relatively shorter time frame than trees grown in natural forests.

## 7 Future trends in research

Research needs include a more intensive search of existing literature reporting on tree growth as influenced by different ground covers to add to the existing database for more reliable estimates on what grasses and legumes produce quality forage and have the least impact on hardwood tree growth and seed production. Few studies have evaluated the impacts of harvesting crops and forages on growth of hardwood trees. Even fewer studies have been done to determine impacts of pruning outside the dormant season. More information is also needed on the impacts of rotational grazing of cattle, sheep and goats in silvopastoral practices on soil properties and tree quality. More evaluation of wood properties and defects are needed as trees mature in established agroforestry practices.

## 8 Where to look for further information

An excellent source of general information on hardwoods that could be managed in an agroforestry practice is the *Silvics of North America – Hardwoods* (Burns and Honkala, 1990). *Agroforestry Systems* (http://www.springer.com/life+sciences/forestry/journal/10457) is the journal most likely to publish new information in response to temperate hardwood trees within the five agroforestry practices.

The Training Manual for Applied Agroforestry Practices (Gold et al., 2015b) and Handbook for Agroforestry Planning and Design (Gold et al., 2015a) prepared by the University of Missouri Center for Agroforestry staff as training guides for their Agroforestry Academy are excellent introductions to the agroforestry practices and their applications. Because most agroforestry practices involve planting open-grown trees and shrubs, *Fruit and Nut Production* by Olcott-Reid and Reid (2007) is an excellent textbook with guidelines for managing the woody component as horticultural crops within these practices. A complementary textbook to this agroforestry book is *North American Agroforestry: an Integrated Science and Practice* edited by Garrett (2009).

There are a number of organizations with annual or biannual meetings that include technical presentations on managing hardwoods for timber production including:

- The Central Hardwood Forest Conference (website changes with each meeting)
- The Association for Temperate Agroforestry (http://www.aftaweb.org)
- The Walnut Council (http://walnutcouncil.org/)
- The Northern Nut Growers Association (https://nutgrowing.org/)
- International Walnut Symposium sponsored by International Society for Horticultural Science (http://www.ishs.org/symposium/)

Centres of expertise include:

- The University of Missouri Center for Agroforestry (http://www.centerforagroforestry.org/)
- Hardwood Tree Improvement and Regeneration Center (https://htirc.org/)
- Dale Bumpers Small Farms Research Center (https://www.ars.usda.gov/southeast-area/booneville-ar/dale-bumpers-small-farms-research-center/)
- National Agroforestry Center (NAC) (https://www.fs.usda.gov/nac/)

# 9 References

Adler, M. J. and Chase, C. A. (2007). Comparison of the allelopathic potential of leguminous summer cover crops: cowpea, sunn hemp, and velvetbean. *HortSci.* 42(2), 289–93.

Alley, J. L., Garrett, H. E., McGraw, R. L., Dwyer, J. P. and Blanche, C. A. (1999). Forage legumes as living mulches for trees in agroforestry practices – preliminary results. *Agrofor. Syst.* 44, 289–91.

Armstrong, J. E., Shigo, A. L., Funk, D. T., McGinnes Jr., E. A. and Smith, D. E. (1981). A macroscopic and microscopic study of compartmentalization and wound closure after mechanical wounding of black walnut trees. *Wood Fiber* 13, 275–91.

Beals, H. O. and Davis, T. C. (1977). *Figure in Wood – an Illustrated Review*. Alabama Agric. Exp. Stn. Bull. #486. Auburn, AL, 5pp. Available at: https://aurora.auburn.edu/handle/11200/2414.

Bedker, P. J., O'Brien, J. G. and Mielke, M. E. (2012). *How to Prune Trees*. NA-FR-01-95. USDA For. Serv. Northeast. Area St. & Private For., 30pp. Available at: https://www.fs.usda.gov/naspf/resources/tree-care-how-prune-trees-english.

Beineke, W. F. (1977). *Corrective Pruning of Black Walnut for Timber Form*. Forestry and Natural Resources Woodland Management FNR-76. Purdue University, West Lafayette, IN. Available at: http://www.agcon.purdue.edu/AgCom/Pubs/FNR/FRN-76.html.

Blizzard, E. M., Kabrick, J. M., Dey, D. C., Larsen, D. R., Pallardy, S. G. and Gwaze, D. P. (2013). Light, canopy closure, and overstory retention in upland Ozark forests. In: *Proceedings of the 15th Biennial Southern Silvicultural Research Conference* (Ed.

Guildin, J. M.), pp. 73–9. e-Gen. Tech. Rep. SRS-GTR-175. USDA For. Serv. South., Asheville, NC. Available at: http://www.fs.usda.gov/treesearch/pubs/43478.

Bragg, D. C. and Stokke, D. D. (1999). Annotated bibliography on 'birdseye' figure grain. Gen. Tech. Rep. NE-263. USDA For. Serv., Radnor, PA, 15pp. Available at: https://www.fs.usda.gov/treesearch/pubs/3795.

Burns, R. M. and Honkala, B. H. (Eds) (1990). *Silvics of North America – Hardwoods* (vol. 2). Agric. Handb. 654. USDA For. Serv., Washington DC. Available at: https://www.na.fs.fed.us/spfo/pubs/silvics_manual/volume_2/vol2_Table_of_contents.htm.

Butin, H. and Shigo, A. L. (1981). *Radial Shakes and 'frost cracks' in Living Oak Trees*. Res. Paper NE-478. USDA For. Serv., Broomall, PA, 21pp. Available at: http://www.fs.usda.gov/treesearch/pubs/15026.

Carpenter, R. D., Sonderman, D. L., Rast, E. D. and Jones, M. J. (1989). *Defects in Hardwood Timber*. Agric. Handb. 678. USDA For. Serv., Washington DC, 88pp. Available at: http://www.fs.usda.gov/treesearch/pubs/13903.

Cassens, D. L. (2004). Factors affecting the quality of walnut lumber and veneer. In: *Black Walnut in a New Century* (Eds. Michler, C. H., Pijut, P. M., Van Sambeek, J. W., Coggeshall, M. V., Siefert, J., Woeste, K., Overton, R. and Ponder Jr., F.), pp. 161–7. Gen. Tech. Rep. NC-243. USDA For. Serv., St. Paul, MN. doi:10.2737/NC-GTR-243. Available at: https://www.fs.usda.gov/treesearch/pubs/14727.

Chen, P. Y. S., Bodkin, R. E. and Van Sambeek, J. W. (1995). Variation in pin knot frequency in black walnut lumber cut from a small provenance/progeny test. In: *Proceedings of the 10th Central Hardwood Forest Conference* (Eds. Gottschalk, K. W. and Fosbroke, S. L. C.), p. 488. Gen. Tech. Rep. NE-197. USDA For. Serv., Broomall, PA. Available at: http://www.fs.usda.gov/treesearch/pubs/12819.

Cunningham, R. A. (1988). Genetic improvement of trees and shrubs used in windbreaks. *Agric. Ecosyst. Environ.* 22/23, 483–98.

Cutter, B. E. (1991). Tree and wood quality in 15-year-old eastern black walnut (*Juglans nigra* L.) grown in agroforestry regimes. In: *Proceedings of the 2nd Conference on Agroforestry in North America* (Ed. Garrett, H. E.), pp. 116–23. Association for Temperate Agroforestry and the University of Missouri, Columbia, MO.

Cutter, B. E. and Garrett, H. E. (1993). Wood quality in alley-cropped eastern black walnut. *Agrofor. Syst.* 22, 25–32.

Cutter, B. E., Phelps, J. E. and Stokke, D. D. (1997). The silviculture-wood quality connection in eastern black walnut. In: *Knowledge for the Future of Black Walnut* (Ed. Van Sambeek, J. W.), pp. 224–30. Gen. Tech. Rep. NC-191. USDA For. Serv. North Central For. Exp. Sta., St. Paul, MN. Available at: https://www.fs.usda.gov/treesearch/pubs/10255.

De Scisciolo, B., Leopold, D. J. and Walton, D. C. (1990). Seasonal patterns of juglone in soil beneath *Juglans nigra* (black walnut) and influence of *J. nigra* on understory vegetation. *J. Chem. Ecol.* 16, 1111–30.

Dey, D. C., Conway, M. R., Garrett, H. E., Hinckley, T. S. and Cox, G. S. 1987. Plant-water relationships and growth of black walnut in a walnut-forage multicropping regime. *For. Sci.* 33(1), 70–80. Available at: http://www.fs.usda.gov/treesearch/pubs/12337 or 12338.

Eisner, N. J., Gilman, E. F. and Grabosky, J. C. (2002). Branch morphology impacts compartmentalization of pruning wounds. *J. Arbor.* 28(2), 99–105.

Gardner, F. P., Pearce, B. B. and Mitchell, R. L. (1985). *Physiology of Crop Plants*. Iowa State University Press, Ames, IA.

Garrett, H. E. (2003). Roles for agroforestry in hardwood management and natural-stand management. In: *Proceedings of the 13th Central Hardwood Forest Conference* (Eds. Van Sambeek, J. W., Dawson, J. O., Ponder Jr., F., Loewenstein, E. F. and Fralish, J. S.). Gen. Tech. Rep. NC-234, pp. 33-43. USDA For. Serv., St. Paul, MN. Available at: http://www.fs.usda.gov/treesearch/pubs/15722.

Garrett, H. E. (Ed.) (2009). *North American Agroforestry: an Integrated Science and Practice*. American Society of Agronomy, Inc., Madison, WI, 379pp.

Garrett, H. E., Neelan, C. and Jones, J. E. (1994). Incidence of yellow-bellied sapsucker damage under walnut agroforestry. In: *Proceedings of the 3rd North American Agroforestry Conference* (Eds. Schultz, R. C. and Anc Colletti, C. P.), pp. 25-8. Association for Temperate Agroforestry and Iowa State University, Ames, IA.

Gingrich, S. F. (1967). Measuring and evaluating stocking and stand density in upland hardwood forests in the central states. *For. Sci.* 13, 38-53.

Gold, M., Cernusca, M. and Hall, M. (Eds) (2015a). *Handbook for Agroforestry Planning and Design*. The Center for Agroforestry, University of Missouri, Columbia, MO, 80pp. Available at: http://www.centerforagroforestry.org/pubs/training/Handbook P&D13.pdf.

Gold, M., Hemmelgarn, H. and Mori, G. (Eds) (2015b). *Training Manual for Applied Agroforestry Practices*. The Center for Agroforestry, University of Missouri, Columbia, MO, 385pp. Available at: http://www.centerforagroforestry.org/pubs/training/.

Hart, J. H. (1982). Variation in inherent decay resistance of black walnut. In: *Black Walnut for the Future*, pp. 12-17. Gen. Tech. Rept. NC-74. USDA For. Serv. N. Cent. For. Exp. Sta., St. Paul, MN. Available at: https://www.fs.usda.gov/treesearch/pubs/10138.

Hart, J. H., Baughan, R. and Jennings, N. E. (1986). Economic impact of decay on black walnut. *North. J. Applied For.* 3, 116-18.

Hoffard, W. H., Heflin, E. L. and Anderson, R. L. (1977). *Economic Impact of Defects in Missouri-grown black walnut*. Impact Survey D-19-77. USDA For. Serv. Northeastern For. Exp. Sta., Broomall, PA.

Jones, J. E., Mueller, R. and Van Sambeek, J. W. (Eds) (1998). *Nut Production Handbook for Eastern Black Walnut*. Southwest Missouri Res. Conserv. & Dev., Inc., Springfield, MO, 147pp. Available at: https://www.fs.usda.gov/treesearch/pubs/10852.

Jose, S. (2009). Agroforestry for ecosystem services and environmental benefits: an overview. *Agrofor. Syst.* 76, 1-10.

Jose, S, and Gillespie, A. R. (1998). Allelopathy in black walnut (*Juglans nigra* L.) alleycropping: II. Effects of juglone on hydroponically grown corn (*Zea mays* L.) and soybean (*Glycine max* L. Merr.) growth and physiology. *Plant Soil* 203, 199-205. doi :10.1023/A:1004353326835.

Jose, S., Benjamin, T., Stall, T., Gillespie, A. R., Hoover, W. L., Mengel, D. B., Seifert, J. R. and Biehle, D. J. (1997). Biology and economics of a black walnut-corn alley cropping system. In: *Knowledge for the Future of Black Walnut* (Ed. Van Sambeek, J. W.), pp. 203-8. Gen. Tech. Rep. NC-191. USDA For. Serv., St. Paul, MN. Available at: https:// www.fs.usda.gov/treesearch/pubs/10255.

Jose, S., Walter, D. and Kumar, B. M. (2017). Ecological considerations in sustainable silvopasture design and management. *Agrofor. Syst.* 93(1), 317-31. doi:10.1007/ s10457-016-0065-2.

Kallenbach, R. (2009). Integrating silvopastures into current forage-livestock systems. In: *Agroforestry Comes of Age: Putting Science into Practice* (Eds. Gold, M. A. and Hall, M. M.), pp. 455-61. Proceedings of the 11th North American Agroforestry

Conference. Association for Temperate Agroforestry and University of Missouri, Columbia, MO.

Kallenbach, R. L., Kerley, M. S. and Bishop-Hurley, G. J. (2006). Forage accumulation, quality and livestock production from an annual ryegrass and cereal rye mixture in a pine-walnut silvopasture. *Agrofor. Syst.* 66, 43–53.

Kesner, A. L. (1986). Characteristics of and management for veneer quality hardwood logs. In: *Guidelines for Managing Immature Appalachian Hardwood Stands Proceedings* (Eds. Smith, H. C. and Eye, M. C.), pp. 22–32. SAF Publ. 86-02. West Virginia Univ. Books, Morgantown, WV. Available at: https://www.fs.usda.gov/treese arch/pubs/50049.

Kormanik, P. P. and Brown, C. L. 1967. Epicormic branching: a real problem in plus tree selection. In: *Proceedings of the Ninth Southern Forest Tree Improvement Conference* (Ed. Tennessee Valley Authority), pp. 69–74. Eastern Tree Seed Laboratory, Macon, GA. Available at: https://rngr.net/publications/tree-improvement-proceedings/s ftic/1967.

Kort, J. (1988). Benefits of windbreaks to field and forage crops. *Agric. Ecosyst. Environ.* 122(23), 165–91.

Krajicek, J. E. (1966). Growing space requirements. In: *Black Walnut Culture*, pp. 47–9. Gen. Tech. Rep. NC-4. USDA For. Serv., St. Paul, MN.

Krajicek, J. E. (1967). Maximum use of minimum acres. In: *Proceedings of the 9th Southern Conference for Forest Tree Improvement* (Ed. Tennessee Valley Authority), pp. 35–7. Eastern Tree Seed Laboratory, Macon, GA. Available at: https://rngr.net/publicat ions/tree-improvement-proceedings/sftic/1967.

Krajicek, J. E., Brinkman, K. A. and Gingrich, S. F. (1961). Crown competition -a measure of density. *For. Sci.* 7, 35–42.

Kremer, R. J. and Kussman, R. D. (2011). Soil quality in a pecan-kura clover alley cropping system in the Midwestern USA. *Agrofor. Syst.* 83, 213–23.

Larsen, M. M. and Schwarz, E. L. (1980). Allelopathic inhibition of black locust, red clover, and black alder by six common herbaceous species. *For. Sci.* 26(3), 511–20.

Larson, P. R. (1963). *Stem Form Development of Forest Trees.* For. Sci. Monograph 5, 42pp. Society of American Foresters, Washington DC.

Law, J. R., Johnson, P. S. and Houf, G. (1994). *A Crown Cover Chart for Oak Savannas.* Technical Brief TB-NC-2. USDA For. Serv. North Cent. For. Exp. Sta., St. Paul, MN. Available at: https://www.fs.usda.gov/treesearch/pubs/10999.

Lehmkuhler, J. W., Kerley, M. S., Garrett, H. E., Cutter, B. E. and McGraw, R. L. (1999). Comparison of continuous and rotational silvopastoral systems for establishing walnut plantations in southwest Missouri, USA. *Agrofor. Syst.* 2-3, 267–79.

Lehmkuhler, J. W., Felton, E. E. D., Schmidt, D. A., Bader, K. J., Moore, A., Huck, M. B., Garrett, H. E. and Kerley, M. S. (2003). Tree protection methods during the silvopastoral systems establishment in midwestern USA: cattle performance and tree damage. *Agrofor. Syst.* 59, 35–42.

Lin, C.-H., McGraw, R. L., George, M. F. and Garrett, H. E. (1999). Shade effects on forage crops with potential in temperate agroforestry practices. *Agrofor. Syst.* 44, 109–19.

Lin, C.-H., McGraw, R. L., George, M. F. and Garrett, H. E. (2001). Nutritive quality and morphological development under partial shade of some forages with agroforestry potential. *Agrofor. Syst.* 53, 1–13.

McKenna, J. R., Geyer, W. A., Woeste, K. E. and Cassens, D. L. (2015). Propagating figured wood in black walnut. *Open Journal of Forestry* 5, 518-25. doi:10.4236/ojf.2015.55045. Available at: https://www.fs.usda.gov/treesearch/pubs/48657.

Meadows, J. S. (1995). Epicormic branches and lumber grade of bottomland oak. In: *Proceedings of the 23rd Annual Hardwood Symposium* (Eds. Lowery, G. and Meyer, D.), pp. 19-25. National Hardwood Lumber Association, Memphis, TN. Available at: https://www.fs.usda.gov/treesearch/pubs/3267.

Miller, L. A., Wray, P. H. and Mize, C. W. (1987). Response of stagnated black walnut to chemical release. *North. J. Applied For.* 4, 93-5.

Miller, G. W., Stringer, J. W. and Mercker, D. C. (2007). *Technical Guide to Crop Tree Release in Hardwood Forests.* SREF-FM-100. Southern Regional Extension Pub SREF-FM-100, 24pp. Available at: https://sref.info/resources/publications/.

Nelson, N. D., Maeglin, R. R. and Walgren, H. E. (1969). Relationship of black walnut wood color to soil properties and site. *Wood Fiber* 1(1), 29-37.

O'Hara, K. L. (2007). Pruning wounds and occlusion: a long-standing conundrum in forestry. *J. For.* 105(3), 131-8. doi:10.1093/jof/105.3.131.

Olcott-Reid, B. and Reid, W. (2007). *Fruit and Nut Production.* Stipes Publishing, L.L.C., Champaign, IL, 597pp.

Pang, K., Van Sambeek, J. W., Navarette-Tindall, N. E., Lin, C.-H., Jose, S. and Garrett, H. E. (2019a). Responses of legumes and grasses to non-, moderate, and dense shade in Missouri, USA. I. Forage yield and its species-level plasticity. *Agrofor. Syst.* 93 (1), 11-24. doi:10.1007s10457-017-0067-8.

Pang, K., Van Sambeek, J. W., Navarette-Tindall, N. E., Lin, C.-H., Jose, S. and Garrett, H. E. (2019b). Responses of legumes and grasses to non-, moderate, and dense shade in Missouri, USA. II. Forage quality and its species-level plasticity. *Agrofor. Syst.* 93 (1), 25-38. doi:10.1007s10457-017-0068-7.

Perkey, A. W., Wilkins, B. L. and Smith, H. C. (1994). *Crop Tree Management in Eastern Hardwoods.* NA-TP-19-93. USDA For. Serv. Northeast. Area State and Private For., Morgantown, WV.

Phelps, J. E. (1989). How management practices influence black walnut wood properties. In: *The Continuing Quest for Quality* (Eds. Phelps, J. E. and McCurdy, D. R.), pp. 21-9. Walnut Council, Indianapolis, IN.

Phelps, J. E. and Chen, P. Y. S. (1989). Lumber and wood properties of plantation-grown and naturally grown black walnut. *For. Prod. J.* 39(2), 58-60.

Ponder Jr., F. (1986). Effect of three weeds on the growth and mycorrhizal infection of black walnut seedlings. *Can. J. Bot.* 64, 1888-92.

Preece, J. E., Navarrete, N. E., Van Sambeek, J. W. and Gaffney, G. R. (1991). An *in vitro* microplant bioassay using clonal white ash to test for tall fescue allelopathy. *Plant Cell Tiss. Org. Cult.* 27, 203-10.

Rast, E. D. 1982. *Photographic Guide to Selected External Defect Indicators and Associate Internal Defects in Northern Red Oak.* Res. Paper NE-511. USDA For. Serv. Northeast. For. Exp. Stn., 20pp. Available at: https://www.fs.usda.gov/treesearch/pubs/15063.

Rast, E. D. and Beaton, J. A. 1985. *Photographic Guide to Selected External Defect Indicators and Associate Internal Defects in Black Cherry.* Res. Paper NE-562. USDA For. Serv. Northeast. For. Exp. Stn., 24pp. Available at: https://www.fs.usda.gov/treesearch/pubs/21736.

Rast, E. D., Beaton, J. A. and Sonderman, D. L. 1988. *Photographic Guide to Selected External Defect Indicators and Associate Internal Defects in Black Walnut.* Res. Paper NE-617. USDA For. Serv. Northeast. For. Exp. Stn., 24pp. Available at: http://www.fs.usda.gov/treesearch/pubs/21802.

Rast, E. D., Beaton, J. A. and Sonderman, D. L. 1989. *Photographic Guide to Selected External Defect Indicators and Associate Internal Defects in White Oak.* Res. Paper NE-628. USDA For. Serv. Northeast. For. Exp. Stn., 24pp. Available at: https://www.fs.usda.gov/treesearch/pubs/21819.

Rast, E. D., Beaton, J. A. and Sonderman, D. L. 1990a. *Photographic Guide to Selected External Defect Indicators and Associate Internal Defects in Yellow-Poplar.* Res. Paper NE-646. USDA For. Serv. Northeast. For. Exp. Stn., 29pp. Available at: https://www.fs.usda.gov/treesearch/pubs/21838.

Rast, E. D., Beaton, J. A. and Sonderman, D. L. 1990b. *Photographic Guide to Selected External Defect Indicators and Associate Internal Defects in Sugar Maple.* Res. Paper NE-647. USDA For. Serv. Northeast. For. Exp. Stn., 35pp. Available at: https://www.fs.usda.gov/treesearch/pubs/21839.

Rast, E. D., Beaton, J. A. and Sonderman, D. L. 1990c. *Photographic Guide to Selected External Defect Indicators and Associate Internal Defects in Yellow Birch.* Res. Paper NE-648. USDA For. Serv. Northeast. For. Exp. Stn., 25pp. Available at: https://www.fs.usda.gov/treesearch/pubs/21840.

Reid, W., Coggeshall, M., Garrett, H. E. and Van Sambeek, J. (2009). *Growing Black Walnut for Nut Production.* Agroforestry in Action AF1011-2009. Univ. Missouri Center for Agrofor., Columbia, MO, 16pp. Available at: http://www.fs.usda.gov/treesearch/pubs/44215.

Rietveld, W. J. (1982). The significance of allelopathy in black walnut cultural systems. In: *Black Walnut for the Future*, pp. 73-86. Gen. Tech. Rept. NC-74. USDA For. Serv. N. Cent. For. Exp. Sta., St. Paul, MN. Available at: https://www.fs.usda.gov/treesearch/pubs/10138.

Rietveld, W. J. (1983). Allelopathic effects of juglone on germination and growth of several herbaceous and woody species. *J. Chem. Ecol.* 9, 295-308. Available at: https://www.fs.usda.gov/treesearch/pubs/19624.

Rink, G. and Van Sambeek, J. W. (1985). Variation among black walnut seedling families in resistance to competition and allelopathy. *Plant and Soil* 88(1), 3-10. Available at: https://www.fs.usda.gov/treesearch/pubs/19625.

Rink, G. and Van Sambeek, J. W. (1987). Variation among four white ash families in response to competition and allelopathy. *For. Ecol. Manage.* 18, 127-34.

Rivest, D. and Vezina, A. (2015). Maize yield patterns on the leeward side of tree windbreaks are site-specific and depend on rainfall conditions in eastern Canada. *Agrofor. Syst.* 89, 237-46.

Rizvi, S. J. H., Tahir, M., Rizvi, V., Kohli, R. K. and Ansari, A. (1999). Interactions in agroforestry systems. *Crit. Rev. Plant Sci.* 18(6), 773-96.

Schlesinger, R. C. (1997). The effects of crowding on black walnut tree growth. In: *Knowledge for the Future of Black Walnut* (Ed. Van Sambeek, J. W.), pp. 139-45. Gen. Tech. Rep. NC-191. USDA For. Serv. North Central For. Exp. Sta., St. Paul, MN. Available at: https://www.fs.usda.gov/treesearch/pubs/10255.

Schlesinger, R. C. (1982). Pruning for quality. In: *Black Walnut for the Future*, pp. 87-91. Gen. Tech. Rept. NC-74. USDA For. Serv. N. Cent. For. Exp. Sta., St. Paul, MN. Available at: https://www.fs.usda.gov/treesearch/pubs/10138.

Schlesinger, R. C. (1986). Pruning Appalachian hardwoods. In: *Proceedings: Guidelines for Managing Immature Appalachian Hardwood Stands* (Eds. Smith, H. C. and Eye, M. E.), pp. 221-7. SAF Publ. 86-02. West Virginia Univ. Books, Morgantown, WV. Available at: https://www.fs.usda.gov/treesearch/pubs/50049.

Schlesinger, R. C. (1988a). Note 3.02 Lateral pruning. In: *Walnut Notes* (Ed. Burde, L.). USDA For. Serv. North Central For. Exp. Sta., 2p. Available at: https://www.fs.usda.gov/treesearch/pubs/11766.

Schlesinger, R. C. (1988b). Note 3.0 First thinning. In: *Walnut Notes* (Ed. Burde, L.). USDA For. Serv. North Central For. Exp. Sta., 2p. Available at: https://www.fs.usda.gov/treesearch/pubs/11767.

Schlesinger, R. C. and Bey, C. F. (1978). Natural improvement in black walnut stem form. In: *Proceedings of the 2nd Central Hardwood Forest Conference* (Ed. Pope, P. E.), pp. 389-400. Purdue University, West Lafayette, IN. Available at: https://www.fs.usda.gov/treesearch/pubs/.

Schlesinger, R. C. and Shigo, A. L. (1989). Pruning central hardwoods. In: *Central Hardwood Notes* (Ed. Hutchinson, J. G.). Note 6.09. USDA For. Serv. North Central For. Exp. Sta., St. Paul, MN. Available at: https://www.fs.usda.gov/treesearch/pubs/11653.

Schlesinger, R. C. and Van Sambeek, J. W. (1986). Ground cover management can revitalize black walnut trees. *North. J. Appl. For.* 3(2), 49-51.

Schlesinger, R. C. and Williams, R. D. (1984). Growth response of black walnut to interplanted trees. *For. Ecol. Manage.* 9, 235-43. Available at: https://www.fs.usda.gov/treesearch/pubs/19623.

Semchenko, M., Lepik, M., Götzenberger, L. and Zobel, K. (2012). Positive effect of shade on plant growth: amelioration of stress or active regulation of growth rate? *J. Ecol.* 100, 459-66. doi:10.1111/j.1365-2745.2011.01936.x.

Shigo, A. L. (1963). *Ring Shake Associated with Sapsucker Injury*. USDA For. Serv. Northeastern For. Exp. Stn., Upper Darby, PA, 10pp. Available at: http://www.fs.usda.gov/treesearch/pubs/3825.

Shigo, A. L. (1984). Compartmentalization: a conceptual framework for understanding how trees grow and defend themselves. *Annu. Rev. Phytopath.* 22(1), 189-214. doi:10.1146/annurev.py.22.090184.001201. Available at: http://www.fs.usda.gov/treesearch/pubs/53382.

Shigo, A. L. (1985). How tree branches are attached to trunks. *Can. J. Bot.* 63, 1391-401.

Shigo, A. L. (1989). *Tree Pruning: a Worldwide Photo Guide*. Shigo and Trees, Durham, NH, 192pp.

Shigo, A. L. (1996). *Tree Basics*. Shigo and Trees, Durham, NH, 40pp.

Shigo, A. L. and Shortle, W. C. (1984). Wound dressings: results of studies over thirteen years. *Arbor. J.* 8, 193-210.

Shigo, A. L., McGinnes Jr., E. A., Funk, D. T. and Rogers, N. (1979). *Internal Defects Associated with Pruned and Nonpruned Branch Stubs in Black Walnut*. Res. Pap. NE-440. USDA For. Serv., Broomall, PA, 27pp. Available at: http://www.treesearch.fs.fed.us/pubs/14974.

Smith, D. E. (1980). Abnormal wood formation following fall and spring injuries in black walnut. *Wood Sci.* 12, 243-51.

Stamps, W. T., Woods, T. L., Linit, M. J. and Garrett, H. E. (2002). Arthropod diversity in alley cropped black walnut (*Juglans nigra* L.) stands in eastern Missouri, USA. *Agrofor. Syst.* 56, 167-75.

Stewart, N. and Dawson, N. (2013). *Forest Thinning – A Landowner's Tool for Healthy Woods*. Ext. Bull. EB-407-2013. University of Maryland, 10pp. Available at: http://extension_umd_edu/woodland.

Stubbs, J. (1986). Hardwood epicormic branching – small knots but large losses. *South. J. Appl. For.* 10, 217-20.

Todhunter, M. N. and Beineke, W. F. (1979). Effect of fescue on black walnut growth. *Tree Planters' Notes* 30(3), 20-3. Available at: https://rngr.net/publications/tpn/30-3/30_3_20_23.pdf/at_download/file.

USDA-FS. (1981). *Quick Reference for Thinning Black Walnut*. USDA For. Serv. North Cent. For. Exp. Stn. and Northeast. Area St. & Private For., St. Paul, MN, 32pp.

USDA-NRCS. (2005). *Land and Water Management Tips for Missouri Landowners*. USDA Natural Resource Conservation Service, Columbia, MO. Available at: https://www.nrcs.usda.gov/wps/portal/nrcs/mo/newsroom/factsheets/.

Van Sambeek, J. W. (2010). Database for estimating tree responses of walnut and other hardwoods to ground cover management practices. *Acta Hortic.* 861, 245-52. Available at: https://www.fs.usda.gov/treesearch/pubs/45876.

Van Sambeek, J. W. (2017). Orchard management using cover crops to improve soil health and pollinator habitat in the Midwestern United States. *The Nutshell (North. Nut Growers Assoc.)* 72(3), 32-45. Available at: https://www.fs.usda.gov/treesearch/pubs/57902.

Van Sambeek, J. W. and Garrett, H. E. (2004). Ground cover management in walnut and other hardwood plantings. In: *Black Walnut in a New Century* (Eds. Michler, C. H., Pijut, P. M., Van Sambeek, J. W., Coggeshall, M. V., Siefert, J., Woeste, K., Overton, R. and Ponder Jr., F.), pp. 85-100. Gen. Tech. Rep. NC-243. USDA For. Serv. North Central Res. Sta., St. Paul, MN. doi:10.2737/NC-GTR-243. Available at: https://www.fs.usda.gov/treesearch/pubs/14714.

Van Sambeek, J. W. and McBride, F. D. (1991). Grass control improves early growth of black walnut more than deep ripping or irrigation. In: *Proceedings of the 2nd Conference on Agroforestry in North America* (Ed. Garrett, H. E.), pp. 42-57. Association for Temperate Agroforestry. https://www.fs.usda.gov/treesearch/pubs/39681.

Van Sambeek, J. W. and Reid, W. (2017). A double row alley-cropping system for establishing nut orchards. *MNGA (Missouri Nut Growers Association) Newsletter* 17(4), 11-14. Available at: https://www.fs.usda.gov/treesearch/pubs/54825.

Van Sambeek, J. W., Ponder Jr., F. and Rietveld, W. J. (1986). Legumes increase growth and alter foliar nutrient levels of black walnut saplings. *For. Ecol. Manage.* 17, 159-67. Available at: https://www.fs.usda.gov/treesearch/pubs/19626.

Van Sambeek, J. W., Schlesinger, R. C., Roth, P. L. and Bocoum, I. (1989). Revitalizing slow-growth black walnut plantings. In: *Proceedings of the 7th Central Hardwood Forest Conference*, pp. 108-14. Gen. Tech. Rep. NC-132. USDA For. Serv. North Central For. Exp. Sta., St. Paul, MN. Available at: https://www.fs.usda.gov/treesearch/pubs/10196.

Von Kiparski, G. R., Lee, L. S. and Gillespie, A. R. (2007). Occurrence and fate of the phytotoxin juglone in alley soils under black walnut trees. *J. Environ. Qual.* 36, 707-17.

White Jr., A. W., Beady, E. R. and Teddars, W. L. (1981). Legumes for supplying nitrogen and studies on legumes in pecan orchards. *Pecan South* 8(4), 24-31.

Willis, R. J. (2000). *Juglans* spp., juglone, and allelopathy. *Allelopathy* 7, 1-55.

# Chapter 3

## New types of products from tropical wood

*Jegatheswaran Ratnasingam, Universiti Putra Malaysia, Malaysia*

## 1 Introduction

In 2015, undisturbed natural tropical forests represented only about 155 million ha, or about 30% of the world's tropical forest areas. The other 70% are constantly being harvested for wood production, and are thus in more urgent need of being managed in a sustainable manner.

According to the International Tropical Timber Organization (ITTO), the term sustainable forest management (SFM) is applicable when the following criteria are met:

  i. SFM is practiced on an operational basis, and not on an experimental scale.
  ii. It should embrace a balanced and comprehensive range of management activities that include working plans, yield prediction and control and other technical requirements.
  iii. It should include the wider political, social and economic criteria without which sustainability is probably unattainable.

http://dx.doi.org/10.19103/AS.2020.0074.10

Against this background, tropical forest management plans have to be effectively implemented to ensure sustainability of the resource (ITTO, 2018).

More than 148 million m³ of industrial round-wood is extracted annually from tropical forests compared with a potential sustainable yield of 134 million m³/year. Almost 52% of this output is used as fuel-wood, while the rest ends up as industrial round-wood and as pulp wood. The FAO (2015) has estimated that, at a global level, there is a widespread tendency to overharvest timber resources. The ratio of sustained yield production versus volume actually harvested varies quite widely:

- the situation in Africa differs by region: in Central Africa, extracted volume exceeds sustainable yield, while in West Africa the threshold of sustainable yield has been largely surpassed and harvested volumes are now more than 200% of sustainable yield in some cases;
- in South America, in general, the levels of sustainable yield and current harvesting are in equilibrium, but most forests in Central America are overharvested, with harvesting up to 10 times the sustainable yield in some cases;
- in Asia and Oceania, forests are generally overharvested, with harvesting exceeding sustainable yield by at least 70%; this does take into account wood harvested during forest clearing and illegal felling.

This provides a context for assessing SFM practices.

## 2 Tropical forest wood products

According to the FAO, industrial round-wood includes wood for the following commodities: saw-logs, veneer logs, pulpwood, other industrial round-wood (used for tanning, distillation, match blocks, poles, piling, posts, pit props and so on) and, in the case of trade, also chips and particles and wood residues. Most of the harvesting of industrial round-wood is undertaken not by large-scale industrial companies, but rather by individuals, families and small operators using a variety of traditional and modern methods.

Production of industrial round-wood increased between 1990 and 2010 in nearly all regions, but the increase, which was slowing down gradually before 1990, slowed considerably between 1980 and 1990. Some of this decline in the rate of increase reflects the demise of the Soviet Union and the associated dramatic reduction in timber production. Russian and other former Soviet Union timber production, in excess of 300 million m³ a year in Soviet times, declined to one-fourth of previous levels (FAO, 2016).

Pulpwood is one component of the industrial round-wood production, and the production of pulpwood from forest ecosystems is often tightly integrated with the production of other solid wood products. Pulpwood is derived from a variety of wood sources, ranging from the harvest of fast-growing young trees in plantations managed specifically for pulp production, to the small or lower-quality stems removed from managed forests to improve forest quality or health, to the shavings, trimmings and other wood produced in the manufacture of sawn wood products. Pulpwood accounts for about a third of the round-wood harvested (including fuel-wood) (IIED, 2017).

In 2015, about 17% of the wood for paper came from primary forests (mostly boreal), 54% from regenerated forest and 29% from plantations (IIED, 2017). This is expected to change as increased output from plantation forests, many of which will mature in the next decade, reaches world markets. The global production of wood pulp has almost tripled in the past 40 years. The rapid increase in wood pulp production over the past four decades was tied to several trends, such as increased population and literacy, leading to increased consumption of paper and paper products and increased use of packaging and packing materials as trade in manufactured consumer goods has grown. The slowdown of the rate of growth in wood pulp production, however, indicates a maturation of these markets and the impact of competing materials. Nevertheless, the International Institute for Environment and Development (2017) predicts continued increased consumption of paper globally, with most of the increase in Asia.

The use of non-wood fiber in pulp supply is expected to remain low and concentrated in Asia. Non-wood materials made up 5.3% of the global pulp in 1990 and 11.7% in 2000, and they are expected to reach 15–18% by 2015 (Pande, 2013). Most of the non-wood fiber is used in 'small-scale pulp mills' (less than 30 tons a day), many of which are currently being closed as a result of poor pollution controls and increasingly stringent standards. This industry has not been well supported by research and development of improved technology or pollution controls (Haberl et al., 2012).

## 3 Plantation forest resources

The establishment of forest plantations can meet a number of needs including:

- carbon fixing;
- the provision of a wood supply source that is an alternative to the natural forest;
- the restoration of degraded land; and
- the generation of income and employment.

About 28 million ha of forest plantations in tropical areas can be considered as forest plantations for wood production (FAO, 2019). These plantations are mainly located in the humid tropical zone. A number of agricultural tree crops can also be considered as commercial plantations (including rubber, coconut and oil palm plantations). It is also currently estimated that more than 500 million ha of degraded land could be planted (or replanted) with trees. However, current rates of forest plantation establishment are only about 1.7 million ha per year (Table 1).

Most large-scale tropical forest plantation establishment (e.g. teak in Asia) has taken place during the last half of this century. The rate of tropical forest plantation establishment worldwide has progressively increased, particularly in South America and Asia (mainly in Brazil, China, Laos and Indonesia, respectively). Experience has shown that the most important factors to consider when establishing a forest plantation are the objectives of management and the vulnerability, in ecological terms, of the site to be planted.

To date, most forest plantations have been established as even-aged monocultures, mainly using exotic rapidly growing species (e.g. Eucalyptus, Acacias and Pines). These species appear to be technically easier to manage and control and more profitable (for wood production) in the short term. The result of this trend has been the restoration of productivity in some forest areas at the cost of a drop in biodiversity and heightened vulnerability to disease, pests and fire. In contrast to single-species plantations, mixed forests have fewer pest control problems, fire is inhibited due to their multi-layered composition, they can restore and maintain soil fertility, and they present a more varied range of development possibilities.

Experience shows that wood yield from forest plantations appears sustainable if species are adapted to the site and if effective management methods are used. In fact, most failures in forest plantation projects occur due to bad species selection or mismatching species and site or the absence of forest

**Table 1** Plantation forest in the tropical region

| | Area of forest plantations for industrial wood production (in million ha) | Annual increase in total forest plantation area (in million ha) | Proportion of total forest plantation area used for industrial wood production (%) |
|---|---|---|---|
| Africa | 2.43 | 0.12 | 52 |
| South America and the Caribbean | 5.97 | 0.23 | 76 |
| Asia and the Pacific | 19.11 | 1.30 | 45 |
| TOTAL | 27.51 | 1.65 | |

*Source*: FAO, (2019).

maintenance and follow-up activities. In such cases, failed forest plantations can actually accelerate soil erosion, water pollution and siltation of watercourses.

Yields from tropical plantations are high, often in the range of 10–30 cubic meters per ha per year for Eucalyptus and Pinus, with some species on favorable soils reaching yields as high as 50–60 cubic meters per ha. Given the amount of research that has gone into improving yield from planted stocks, these yields are likely to continue to increase. Because of high yields and increasing area, plantations provide a continuously increasing portion of the world's timber supply. According to the FAO (2001), plantations were only about 5% of the global forest cover in 2000, yet provided some 35% of global round-wood supply, an amount anticipated to increase to 44% by 2020.

Tropical forest plantation establishment and management techniques are currently well known for many species. Measures to improve tropical forest plantations have yielded notable results, but socioeconomic and technical constraints (e.g. low manpower availability and difficulties in expanding the result of experiments to large areas) have led to a greater interest in the development of intensive mechanized methods (Ratnasingam and Awang, 2013).

Given current forest product prices, forest plantations nearly always turn out to be too costly to be economically viable in the short term. Furthermore, when considering that the revenue from plantation investments arises in the long term, they are constantly exposed to various risks, such as collapses in prices, natural disasters and political instability. Another factor that has to be taken into account in the appraisal of tropical forest plantation programs is their impact on the landless and poor people who are directly competing to use the land. In some cases, tropical forest plantations have led to the eviction of traditional users, upset existing systems of harvesting and extraction for local use and fostered serious social conflict. Thus, it is necessary to consider the wider impacts of such developments, but little concrete information about such issues are usually available. It must be recognized that forest plantations are becoming increasingly important throughout the world as the primary source of wood material. It is especially relevant when large quantities of uniform quality fiber are required for pulp and paper production. This is to be anticipated because the pulp and paper industry is capital intensive, and the cost contribution of the raw material is a fraction of the total investment incurred. Therefore, a sustainable supply of raw material is crucial to maintain high-production capacities (>90%) in such mills, to ensure its economic viability (Ratnasingam and Awang, 2013).

In the final analysis, however, the establishment of forest plantations in the tropics should be governed by very strict guidelines and standards related to the design, planning and maintenance of these plantations. Factors such as land availability, species-site matching, species-climate matching, site preparation, forest plantation health, pests and diseases risks, environmental values, biodiversity conservation, social benefits, and the proper harvesting and extraction techniques should be taken into consideration before establishing the forest plantation. There are many instances in the tropics when forest plantations schemes failed because of poor adherences to the above factors (Ratnasingam and Awang, 2013).

## 4 Non-wood forest products: bamboo and rattan

Many non-wood forest products are of importance to people in every forest ecosystem throughout the world. These products contribute directly to the livelihoods of an estimated 500 million people worldwide, either directly or indirectly. These products include foods, medicinal products, dyes, minerals, latex and ornamentals among others (Pande, 2013).

### 4.1 Bamboo

There are approximately 1200 species of woody and herbaceous bamboos, the former being most important from the socioeconomic perspective. Many woody bamboos grow quickly and are highly productive (Dransfield and Widjaja, 1995). Annual productivity values range between 10-20 tons per ha per year, and bamboo stands may achieve a total standing biomass that is comparable to some tree crops (of the order of 20-150 tons per ha) (Hunter and Junqi, 2002). It is estimated that bamboo makes up about 20-25% of the terrestrial biomass in the tropics and subtropics (Bansal and Zoolagud, 2002). A substantial amount of bamboo timber comes from plantations, although natural forests are also important.

Bamboos are multipurpose crops, with more than 1500 documented uses. As a construction material, bamboo is widely used in all parts of the world where it grows, and because of its high strength-weight ratio, bamboo is the scaffolding material of choice across much of Asia. The tubular structure of the plant is optimally 'engineered' for strength at minimum weight. In many places, its use is restricted almost exclusively for low-cost housing, usually built by the owners themselves. For this and for other reasons, bamboo is often regarded as the 'poor man's timber' and used as a temporary solution to be replaced as soon as improved economic conditions allow. However, architects' interest in working with bamboo has also led to this becoming a common building material for the wealthy. Modern manufacturing techniques allow the use of bamboo in timber-based industries to produce flooring, board products, laminates and furniture (Bansal and Zoolagud, 2002).

Although reported figures on the area of bamboo forests are inconsistent, it is widely accepted that China is the richest country in the world in terms of bamboo resources. China's bamboo forests cover an estimated area of

44 000-70 000 km$^2$, mostly of *Phyllostachys* and *Dendrocalamus* species. Their standing biomass is estimated to be more than 96 million tons (Feng et al., 2001). Asia ranks first in bamboo production, and Latin America is second. It is estimated that bamboo in Latin America covers close to 110 000 km$^2$ (FAO, 2015).

Worldwide, domestic trade and subsistence use of bamboo is estimated to be worth $8-14 billion per year. Global export of bamboo generates another $2.7 billion (INBAR, 2017). Bamboo is increasingly being used as a substitute for wood in pulp and paper manufacturing, and currently India uses about 3 million tons of bamboo per year in pulp manufacture and China about 1 million tons, although China is set to increase the use of bamboo for paper to a target of 5 million tons per year. In many countries in Asia, Africa, and Central and South America, bamboo products are used domestically and can be very significant in both household and local economies (FAO, 2001).

### 4.2 Rattan

Rattan is a scaly, fruited climbing palm that needs tall trees for support. There are around 600 different species of rattan, belonging to 13 genera, the largest of which is *Calamus*, with some 370 species (Sunderland and Dransfield, 2000). It is estimated that only 20% of the known rattan species are of any commercial value. The most important product of rattan palms is cane from the stem stripped of leaf sheaths. This stem is solid, strong and uniform, yet highly flexible. The canes are used either in whole or round form, especially for furniture frames, or split, peeled, or cored for matting and basketry. Rattans require considerable treatment, including boiling and scraping to remove resins and dipping in insecticides and fungicides prior to drying. The range of indigenous uses of rattan canes is vast, from bridges to baskets, fish traps to furniture, crossbow strings to yam ties.

Rattans are almost exclusively harvested from the wild tropical forests of South and Southeast Asia, parts of the South Pacific (particularly Papua New Guinea) and West Africa. Much of the world's stock of rattan grows in over 5 million ha of forest in Indonesia. Other Southeast Asian countries, such as the Philippines and Laos, have less rattan but have been relatively self-sufficient due to the appropriate size of their processing sector. No rattans grow naturally elsewhere, and even in these locations deforestation can lead to local extinction of rattans due to their dependence on mature forests.

In the last 20 years, the international trade in rattan has undergone rapid expansion. The trade is dominated by Southeast Asia, and by the late 1980s the combined annual value of exports of Indonesia, the Philippines, Thailand and Malaysia had risen to almost $400 million, with Indonesia accounting for 50% of this trade (Sunderland and Dransfield, 2000). Worldwide, over 700 million people trade in or use rattan. Domestic trade and subsistence use of rattan are estimated to be worth $2.5 billion per year. Global exports of rattan generate another $4 billion (INBAR, 2017).

## 4.3 Drivers of change in bamboo and rattan products

Traditionally, bamboo was used domestically and supplies were extracted based on local requirements. Contemporary additional applications of bamboo have propelled it into new domestic and international markets, increasing profits and income for many participants in the sector. Bamboo generates substantial export income for several countries, such as China ($329 million in 2015) and the Philippines ($241 million in 2015) (INBAR, 2017).

Indonesia has a clear advantage over other countries, with its overwhelming supply of wild and cultivated rattan (80% of the world's raw material), and rattan contributes about $500 million to Indonesia's foreign exchange and is an important vehicle for rural development. It also raises the value of standing forests, as rattan is the most valuable of the non-wood forest products in the country, earning 90% of total export earnings from such products (INBAR, 2017).

Most bamboo and rattan processing countries are facing a shortage of raw material. The primary cause of this trend ranges from overharvesting and conversion of bamboo and rattan forestland to settled agriculture or shifting cultivation. Restoring productive agricultural land to bamboo and rattan production is often difficult, as concerns for food security is usually more pressing and attracts more attention (INBAR, 2017).

## 5 Plantation timber product development and wood species

Over the last five decades, the wood products sector throughout the tropical region has flourished on imported knowledge and technology, especially from the United States, Europe and Japan, in terms of wood products development. In fact, it is obvious that almost all existing wood products manufacturing have been developed along such a technology pull strategy (FPL, 2001). Sawn timber, wood treatment, wood drying, thermal treatment of wood, mechanical wood processing, wood-based panel manufacturing, joinery and carpentry products, furniture, engineered wood composites such as glulam, cross laminated timber (CLT) and so on, nano-cellulose products and bioenergy are among the many products driven primarily by the availability of technology and know-how, as well as market demand.

As reflected in Fig. 1, the necessary inputs for the development of new wood products in the tropical region have been the amply available factor inputs, especially raw material, workforce and technology. Hence, the wood products

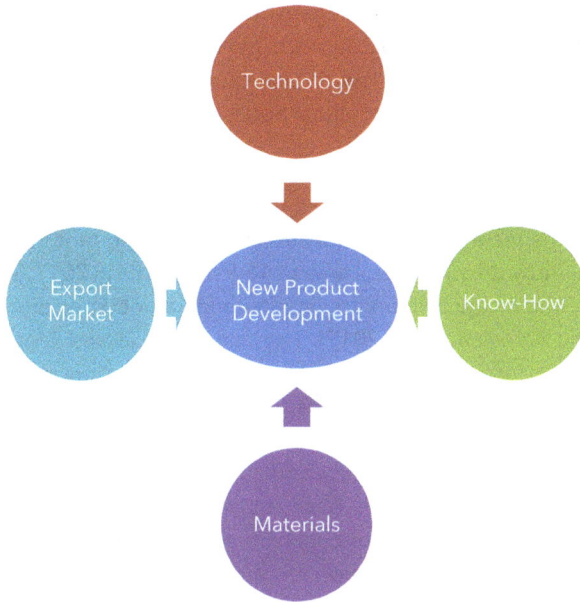

**Figure 1** New product development driving factors.

industry in the tropical region has been registering steady annual double-digit growth over the years (Amar et al., 2005). Since the late 1990s, when China emerged as a leading technical solution provider for the wood industry, the pace of the wood industry's growth in the tropical region accelerated further due to the newly found cost competitiveness (FAO, 2016).

To a large extent, the pull-factor for the rapid development of wood products in the tropical region has been fueled by availability of the diverse supply of raw materials:

1. Plantation wood resources;
2. Plantation biomass and waste wood; and
3. Agricultural biomass and waste.

In the following subsections, these resources will be discussed in more detail.

## 5.1 Rubberwood

*Hevea brasiliensis* commonly referred to as rubberwood or heveawood has been widely accepted in Asia. Despite many new fast-growing tree species being introduced as potential substitutes for timber from the natural forests, the demand for rubberwood is still growing steadily. Rubberwood is popular as a raw material for manufacturing wood composites, such as particleboard and medium-density fiberboard (Ratnasingam et al., 2012). Rubberwood composites are frequently used in many applications, especially for furniture and partitioning (Fig. 2). However, due to their low natural durability, this resource has found limited use in engineering and structural applications (Ratnasingam et al., 2012).

### 5.2 Eucalyptus

The Eucalyptus species populate the second most important industrial forest plantations worldwide, with a total area of 18.9 million ha in 2017. In the tropics, Eucalyptus is one of the primary forest plantation wood species, especially in Asia and Latin America (FAO, 2019). The fast growing and high wood-fiber quality makes Eucalyptus an interesting raw material for the pulp and paper industry. Because of its potential industrial use, large quantities of Eucalyptus bark are also generated annually and burnt for energy production. When compared to other biomass products and considering its fibrous structure, this is not their optimal use, primarily because of its relatively low heating value. Therefore, a significant quantity of Eucalyptus bark should be readily available and exploited for the manufacture of new and higher-value-added products.

In terms of the wood quality, Eucalyptus wood is suitable not only for furniture, but its application has also been extended to the manufacture of builders, joinery and carpentry (BJC), flooring and some engineered composites. Trials on using Eucalyptus wood for structural boards, such as glulam and cross-laminated timber (CLT) is also being explored, but no major commercial success has been reported so far (Brandner et al., 2016) (Fig. 3).

**Figure 2** Finger jointed rubberwood.

**Figure 3** Natural wood veneer from Eucalyptus.

## 5.3 Acacia

*Acacia mangium* is a subfamily species within the family of Mimosoideae, a leguminous tree. This wood species was vastly planted throughout many parts of Asia since the 1980s, but due to its susceptibility to heart rots, interest was shifted towards hybridization of these *Acacia* (Bakri et al., 2018). In recent years, plantations of Acacia Hybrid (AH) (mixture of *Acacia auriculiformis* and *Acacia mangium* (AA and AM) and second-generation AM, which is known to grow faster, with high resistance to heart rot, possessing a straighter bole structure, and having less tapering have been promoted successfully. Further, Acacias have also been exploited to the full, especially in terms of its life-cycle assessment (LCA). In this context, the wood has been applied in a wide variety of applications, including structural lumber, veneer, plywood, wood-based composites, furniture, flooring and so on. In fact, the success of Acacia as a raw material for the wood products industry is clearly demonstrated by the Vietnamese wood products sector, in which Acacia accounts for almost 60% of all wood resources used (Bakri et al., 2018) (Fig. 4).

## 5.4 Other plantation wood species

The other common plantation wood species in the tropical region include Teak (*Tectona grandis*), Pinus spp., *Gmelina arborea*, *Albizia falcataria* and *Paulownia* spp. These plantation tree species are being aggressively cultivated in many parts of tropical Asia and Latin America, as these plantations are expected to offset the shortfall in the supply of wood resources from the natural forests, apart from increasing the forest cover in the respective countries. These wood resources are being exploited for a variety of purposes, but not all have been successful for its intended purposes (Ratnasingam and Awang, 2013; FAO, 2019).

**Figure 4** Acacia wood finger joint.

# 6 Plantation biomass and waste: oil palm and kenaf

## 6.1 Oil palm

The oil palm industry produces huge quantities of biomass waste, both in the field and the oil palm mills. The waste from mills consists of pressed fruit fibers (PFF), empty fruit bunch (EFB), oil palm shell (OPS), palm oil mill effluent (POME), while the wastes from the plantation fields comprise of the oil palm trunks (OPT) and oil palm fronds (OPF), especially during the replanting of the old mature trees.

These renewable biomass sources can be used for the development of bio-composites, power generation, paper production, construction board fillers, solid wood, mulching and soil conditioning as well as many other uses. Its ample availability, low price, reasonable performance and its biodegradable nature are among the factors that act as catalysts to promote its use for the manufacture of value-added products. Unfortunately, only 10% of the available oil palm biomass has found commercial applications at this present time, for bio-composite-based industries, industrial raw materials, fertilizers, animal feeds, chemical derivatives and others. Much of this biomass wastes which is not presently exploited contributes to severe environmental problems, in relation to its greenhouse gas (GHG) emission as a result of its composition. Previous research on biomass and other agricultural waste has shown potential in its use for the production of various types of value-added products, such as medium-density panel, chip board, thermoset composite and thermoplastic, nano-bio-composite, pulp and paper manufacture (Ratnasingam and Milton, 2011).

Through intensive research and development activities since the late 1980s, the world's oil palm-producing countries have been able to commercialize a variety of biomass-based products. In this context, the use of oil palm biomass for various types of value-added products through physical, biological and chemical technology and innovation is slowly evolving.

In recent years, the biomass wastes from the oil palm industry, which include the trunk, empty fruit bunch, leaf, meso-carp fiber and so on, have been converted into various bio-composite products. (Fig. 5). Conventional composites, such as veneer, plywood, compressed lumber, sandwich panel, particleboard and fiberboard are used in some structural and non-structural product applications, including panels for internal closure purposes and panels for outdoor use in furniture and multi-building support structures (Ratnasingam and Ioras, 2017). In fact, one of the most promising research into the commercial use of oil palm trunk has been the production of sawn lumber for the manufacture of sandwich and insulation boards, but its full commercial realization has been thwarted by its prohibitively high technology cost from Germany (Ratnasingam and Ioras, 2017).

## 6.2 Kenaf

Kenaf fiber is made from the bast (outer) and core (inner) fibers of the kenaf plant (*Hibiscus Cannabinus* L.). Kenaf fiber is native to Africa and Asia and is one

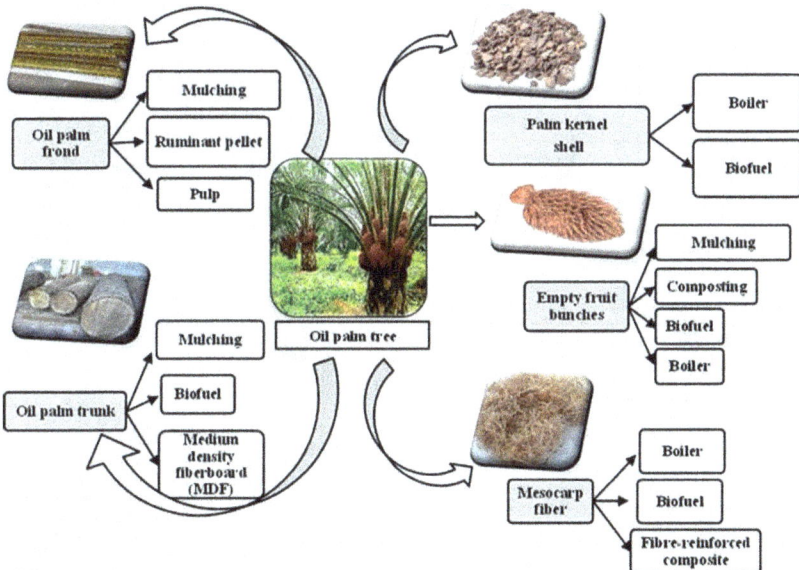

**Figure 5** Oil palm biomass utilization.

of the most widely cultivated natural fibers (Akil et al., 2011). It is a herbaceous annual plant that is cultivated commercially in many countries throughout the world, in a variety of weather conditions for a variety of uses. Kenaf can grow up to a height of 2.4-6 m at an average of 150 days (Amar et al., 2005). The leading countries in kenaf production in the world are India and China (Stokke et al., 2014). From an agronomic perspective, kenaf has advantages with regard to their resistance to climatic extremes, pests and diseases (Feng et al., 2001).

Kenaf is mostly known for its fibers, as these are the largest fractions of the kenaf plant. In a kenaf stalk, there are two types of fibers, the inner and the outer fibers namely the core and bast, respectively. The characteristic of having two different regions of fibers in a stalk is similar to flax straw, jute and hemp. These two fibers are so versatile because they can be used for many different purposes, such as to be converted into pulp, paper, cardboard, panels, traditional cordages, absorbent agent, packing materials, soil-less potting mixes, grass and flower mats and natural fuels.

One of the prevalent used of kenaf fibers is in the manufacture of bio-composites. Bio-composite is defined as a material formed from a combination of matrix (resin) and natural fibers (wood or non-wood) (filler) which are obtained from plants. Usually, in the fabrication of bio-composites, the important parameters that need to be taken seriously are the pressing pressure and pressing temperature.

Bio-composites can be divided into two broad groups: (1) fully biodegradable and (2) partially biodegradable. In the case of the kenaf fibers, it is often used as a reinforcement in the polymer matrix, in the form of particulate fillers, short, continuous fiber, non-woven and woven fabrics. Kenaf bio-composites have found commercial applications in the automotive, furniture, wood-plastic and even in armored panels manufacturing industry (Holbery and Houston, 2006). However, the most successful application of up-to-date kenaf fibers is in the production of automotive components, which has a number of requirements that include:

a) Consistent, good quality and sustainable materials;
b) Wettable and interactive bio-composites;
c) Durable bio-composites;
d) Odourless bio-composites;
e) Fully 'green' bio-composites; and
f) High-performance and competitive bio-composites (Fig. 6).

Consistent, good quality and sustainable materials: One of the main requirements of kenaf utilization in automotive is the sustainable supply of good and consistent quality of industrial grade fibers. The chemical and physical morphology of the kenaf fibers, their cell wall growth patterns and thickness,

**Figure 6** Bio-composite products from Kenaf.

dimensions and shapes of the cells, cross-sectional shapes, distinctiveness of lumens and so on, besides their chemical compositions (cellulose, hemi-cellulose and lignin contents), have profound effect on the properties of the fibers (Cao et al., 2006). The variability in the fiber properties can lead to variability of mechanical properties of the resulting bio-composites.

Wettable and interactive bio-composites: One of the main challenges in producing high-performance bio-composites is to improve the degree of interaction or interfacial bonding between hygroscopic bio-fibers and the hydrophobic or apolar nature of various types of polymer matrices, that is, thermoplastic and thermosetting resins (Ansell, 2015). A clear disadvantage of this apolar character for composite applications is its limited wettability, as well as poor interfacial bonding with reinforcing fibers. Thus a critical assessment of the interfacial bond is needed for a successful design of the final component.

Durable bio-composites: The presence of hygroscopic bio-fibers (due to high content of hydroxyl groups) in bio-composites resulted in composite products to absorb moisture over time, causing functional and durability concerns (Cao et al., 2006). In addition, this could give rise to odor and smell of the automotive components. This is extremely crucial in the context of the tropics, which experience high relative humidity throughout the year. More in-depth research is thus required to find the cost-effective means of increasing the moisture resistance and durability of bio-composites.

Odorless bio-composites: One of the problems in combining kenaf with a polymeric matrix, such as polypropylene (PP), is the poor thermal stability, which results in degradation of the fiber at the common processing temperatures of the composites. This process leads to the darkening of the material and formation of low molecular weight compounds that produce undesirable odor. This issue is very pertinent in the context of market acceptance and consumer perception of the final product. It is therefore, important to minimize the effects of fiber chemical treatments and chemical modification on the polymer matrices, so as to reduce the emission of odorous volatile organic compounds (VOC).

Fully 'green' bio-composites: The challenge for the automotive industry is to produce lighter, inexpensive, environmentally sustainable vehicles that are safe, attractive, energy efficient and economical to operate. The social push and increasing environmental awareness of consumers in the developed world have provided the ground for increasing the use of the so-called 'green' materials in the automotive sector (Ansell, 2015; Davim and Aguilera, 2017). The current trend is toward using fully green bio-composites, which involve the combination of kenaf fibers with biodegradable polymers. Several natural resin systems (both thermoset and thermoplastic) have been considered as candidates for the automotive industry. Thermoplastic bio-resins, such as polylactic acid (PLA), polyalkanoates (PHA) and poly butylene succinate (PBS) have potential applications because of their biodegradable nature. However, market penetration of these bio-resins in the automotive industry is presently restricted by its prohibitive cost. In order to achieve the wide use of bio-resins, the controlled synthesis of resins with a wide range of properties to suit different applications is required, so as to ensure economies of scale. Thus extensive R&D efforts are being directed toward lowering the cost of production and diversifying the properties of these bio-resins, in order to make them competitive in comparison to conventional plastics.

Realizing high performance and competitive bio-composites: In order to be able to compete with glass-fiber-reinforced composites and gradually become an important replacement material for automotive components, the performance of bio-composites needs to be uplifted in several ways. One of the key aspects that need to be focused is finding means of having highly filled bio-composites with high aspect ratio (length to diameter ratio), especially of the kenaf fibers. From a practical point of view, kenaf fibers have dimensions that are limited by their anatomical restrictions (Holbery and Houston, 2006). In terms of length, it has been found that many natural fibers have average values that are sufficient to be considered as long and continuous reinforcement for the fabrication of bio-composites (Cao et al., 2006). The use of conventional melt-processing methods for the production of short fiber-reinforced compounds, such as extruders will not be acceptable, since it will result in serious degradation of fiber length. High fiber-aspect ratio is very crucial in achieving high performance bio-composites. In light of this, considerable R & D efforts are now being undertaken in finding means of producing bio-composites with high fiber loadings and high fiber-aspect ratios, regardless of the fact that both variables will affect the rheology and limit the processing of such composites. Inevitability, fundamental research is essential to achieve a balance between mechanical performance and processing properties of such bio-composites.

## 7 Plantation biomass and waste: sugarcane and rice

Although world prices for sugar have shown spectacular variations since 1973, the long-term outlook is likely to be a very gradual increase in the price for sugar (Klass, 2011). This uncertain prospect explains, to a large degree, the renewed interest in the by-products of the sugarcane industry, which has rapidly developed in the last ten years. The optimal use of these by-products can provide a significant support to the sugarcane industry, although it could not, by itself, completely redress the difficult market situation that sugar is presently experiencing. The four main by-products of the sugarcane industry are cane tops, bagasse, filter muds and molasses. However, the most important of these by-products is bagasse.

Bagasse is used for the generation of steam and power required to operate the sugar factory. A typical factory producing a ton of cane sugar requires about 35 kWh and 450 kg of exhaust steam. Much progress has been achieved lately with the technology for the continuous operation of pans, crystallizers and centrifuges. Inevitably, efficiency has improved and presently a modern raw sugar factory can now operate with 30 kWh and 300 kg of exhaust steam to produce one ton of cane sugar. Such a factory can save up to 50% of the bagasse it produces, and this bagasse can then be used to produce electricity for the grid, or can be diverted as a raw material for the production of paper, board, furfural and so on (Klass, 2011).

The more straightforward solution is to produce electricity from bagasse through a high-pressure boiler and condensing turbo-alternator. This solution has found favor in a number of sugarcane-producing countries, such as Australia and Mauritius. In these countries with modern equipment nearly 450 kWh can be generated from one ton of mill-run bagasse. A typical example of this is to operate a power generation plant using mill-run bagasse on a year-round basis. If the bagasse is priced at US$ 15 per ton under prevailing market condition, the cost of electricity generated is approximately US cents 6 to 8 per kWh, which is very competitive for most Third World countries (Seifert, 2014).

To be economical, this will require bagasse storage through various methods such as, dry and wet bulk storage, bale storage and pelleting, to ensure consistent supply during the intercrop season. Such an effort of generating electricity from bagasse is undoubtedly the easiest and best utilization of this by-product for most sugarcane producing Third World countries.

The production of particle board from bagasse is a well-proven technology, but it has to compete with conventional wood-based particleboard, plywood and fiberboard. Its main difficulty is the high cost of imported synthetic resins which serve as a binder to the bagasse fibers composing the board. Further, the board's optimal thickness is often less than 15 mm, and its market is limited to indoor applications for inner partitions and furniture. In the last few years, however, a process has been developed in the Federal Republic of Germany whereby Portland cement replaces the urea formaldehyde (UF) resin, which enables this cement-bonded particle board to be used for exterior walls, roofing and so on. Such a product has a higher market appeal and allows the product to be extended to exterior applications. Nevertheless, it must be noted that the bagasse utilized should not contain more than 0.5% of sugar on a bone-dry basis, as higher sugar contents can retard cement-setting, resulting in poor-quality boards (Shmulsky and Jones, 2019).

Good-quality wrapping and magazine paper can be produced with a high percentage of de-pithed bagasse as raw material. The availability of a sizable market, sufficient amount of bagasse and ample supply of quality industrial water are the usual constraints faced in many instances, apart from the high capital intensity of establishing paper mills and the necessity to handle polluting effluents. Up to now, however, the production of newsprint from bagasse has proved difficult and uneconomic. Nevertheless, there are constant advances in pulp and paper technology, and eventually, bagasse newsprint may become feasible within the next ten years, especially if mixed with a fair percentage of waste paper. Also the production of magazine or note paper on a small scale has been investigated by Shmulsky and Jones (2019), and the experiences gained in India seem to confirm the feasibility of establishing mills to produce as little as 15 tons of bleached bagasse-based pulp per 24 hours.

The process generally favored for the production of bagasse pulp is the kraft process using sodium sulfate. The actual sulfate cooking liquor contains a 4:1 mixture of caustic soda and sodium sulphide. Typical yield on bone dry de-pithed bagasse, to be expected with the kraft cooking process is 48% for the final bleached slush pulp (Fig. 7).

Against this background, the production of pulp and paper from bagasse is not advisable as the main use of sugarcane by-products by Third World countries, unless very favorable local conditions exist. It must be enunciated that pulp and paper manufacturing requires relatively demanding technology, which is best approached after gaining experience with simpler bagasse processing, involved in ventures such as electricity generation or particle board manufacturing.

**Figure 7** Sugarcane paper and briquette.

## 7.1 Rice husk

Rice husk is a major by-product of paddy processing in countries, such as Vietnam, Thailand and China. The rice husk accounts for about one-fifth of the paddy produced, on weight basis. Estimated annual production of rice husk in these countries is approximately 48 metric tons annually, and therefore, economic utilization of this by-product is highly desirable.

The main constituents of rice husk are cellulose, pentosan, lignin and silica. All these constituents have valuable industrial applications. Pentosan is considered as a good source for furfural production. Amorphous silica is a very good source for the preparation of pure silicon and a number of other silicon compounds, such as silicon carbide, silicon nitride, cement, ceramic and other silicate materials. Cellulose is the raw material for pulp and paper making.

Rice husk with a bulk density of 100–160 kg/m³ has a thermal conductivity of about 0.0359 W/m°C, rendering it excellent insulation properties, comparable with other commonly used insulating materials. Further, it has energy content between 11.9 MJ/kg and 13 MJ/kg at 14% moisture content. In terms of nutrients, it has less than 10% total digestible nutrients (Ansell, 2015). Rice husks have found applications as:

- Fuel;
- Fiberboards;
- Ash and pure silica;
- Production of furfural; and
- Animal feed.

Traditionally, rice husk has been mostly used as a fuel to provide energy for rice mill operation. However, most of the existing furnace designs converted only about half of the available energy in the husk, while the remaining heat potential remains in its fixed carbon portion, which creates a disposal problem.

For binder-less board production, the powdered rice husk (0.25-0.39 mm) is mixed with 5-8% (by weight) of concentrated sulphuric acid, and then dried in the sun. The dried powder is eventually pressed at 60-70 kg/cm² for 20-25 min at 165°C or 12-15 min at 175-200°C to form the fiberboard. On the other hand, fiberboards using binder can be made by mixing rice hull powder (0.78-1.68 mm at 4-6% moisture content) with 7-8% (by weight) phenol-formaldehyde (PF) resin, dried in the sun, and molded in a hydraulic press into 1 mm thick board by pressing for 1 min at a temperature of 175-200°C. (Fig. 8).

Rice husk-ash is a unique source of high-grade amorphous silica. The silica present in rice husk, of biogenic origin, is inherently amorphous. Normally, controlled combustion below 700°C yields white ash with amorphous silica. On the contrary, amorphous silica obtained from rice husk is chemically active and hence a very useful product. At higher temperatures, it undergoes a phase change resulting into crystalline forms of silica (Csanàdy et al., 2019).

Furfural from rice husk can be prepared by acid-hydrolysis using sulphuric acid (of less than 0.5M) and super-heated water at 185°C. Furfural products can replace formaldehyde in phenol-formaldehyde (PF) resins. It can form chemically inert materials useful for a number of products, including the manufacture of corrosion-resistant pipes. Furfural resin is also used to dissolve the undesirable constituents of lubricating oil.

Rice husk forms the major part of the concentrate fed to cattle and goats in India. However, due to its low total digestible nutrient level, the level of husk in the concentrate feed should not exceed more than 15% for cattle and 25% for goat. To improve the digestibility of the rice husk, treatment with alkali (NaOH) to facilitate fermentation has been suggested, but has not yet found widespread application.

**Figure 8** Rice husk wood plastic composite.

# 8 Developments in wood products

The wood products industry, regarded as a labor-intensive, cottage-based industry, was highly dependent on high-quality solid wood resources as its feedstock. Through technological advancements and market innovations, this traditional industry has been transformed into a relatively advanced industry that could convert and utilize a variety of raw materials, both wood and non-wood fiber resources into value-added products (Ratnasingam and Milton, 2011).

The wood products industry in many countries throughout the tropical region has emerged as an important socioeconomic sector, both as a foreign exchange earner and also as an employment provider, especially to rural communities. In this respect, primary wood products such as sawn timber, veneer (both peeled and sliced), plywood, laminated veneer lumber, and also the secondary or remanufactured wood products which include moldings, builders' carpentry and joinery and furniture, are all being increasingly manufactured in many tropical countries, both for domestic and also for export markets (FPL, 2001). Products from harvesting biomass, mill waste and off-cuts are also finding widespread applications in the manufacture of wood-based panels, especially particleboard, medium-density fiberboard (MDF), and lately, the manufacture of oriented strand board (OSB). These wood products have all been developed several decades ago, and any new developments is often associated with the use of a new raw material resource, rather than any major improvements in product design and structure (Shmulsky and Jones, 2019). (Fig. 9).

With a push toward the green consumerism and economy, in line with the sustainable development goals (SDGs) initiatives, wood and wood products are growing in importance both for non-structural and structural applications. Its low carbon foot-print, sustainable resource status as well as it being an integral part of the circular economy, are among the major factors that have boosted the interest toward wood and wood products throughout the world (Milner, 2009).

In the case of the tropics, the extensive use of wood and wood products have primarily been confined to non-structural applications, with limited use for structural applications due to the concerns associated with its durability in the

**Figure 9** Types of traditional wood products.

harsh tropical environment and also its perceived low fire-resistance. Hence, it is not surprising that engineered wood products development, and its ensuing applications in the tropical region, has been relatively slow (Milner, 2009). Inevitably, most of the engineered wood products available are very much in the early stages of market development.

Cross-laminated timber (CLT) is a wood panel system that is gaining popularity in the United States and the rest of the world, after being widely adopted in Europe. CLT is the basis of the tall wood building movement, as the material's high strength, dimensional stability and rigidity allow it to be used in mid- and high-rise construction (Brandner et al., 2016). CLT panels are made up of layers of lumber boards (usually three, five or seven) stacked crosswise at 90-degree angles and glued into place. (Fig. 10). The panels can be manufactured at custom dimensions, though transportation restrictions dictate their length.

> Applications for CLT include floors, walls and roofing. The panels' ability to resist high racking and compressive forces makes them especially cost-effective for multistory and long-span diaphragm applications. Some specifiers, architects and building contractors view CLT as interchangeable with other wood products and building systems. Like other mass timber products, CLT can be used in hybrid applications with materials such as concrete and steel. It can also be used as a prefabricated building component, accelerating construction timelines, which in turn, reduces the overall building construction cost.

Several factors that contribute to a growing market for CLT and tall wood construction, among others, include: advances in wood connectors, the development of hybrid materials and building systems, the successful commercialization of CLT and growth in off-site fabrication. The alternating grain directions in the CLT's panel improve its dimensional stability. The individual lumber boards used in CLT, typically vary in thickness from 15 mm to 50 mm, and in width from 60 mm to 240 mm. Finger joints and structural adhesives are used to connect these boards.

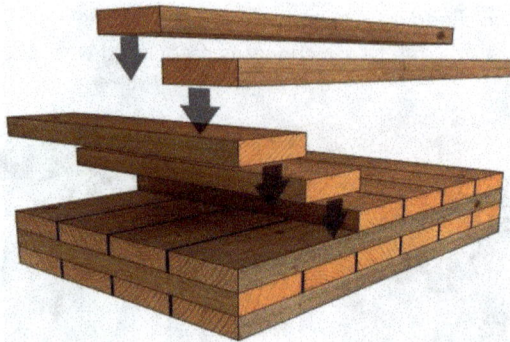

**Figure 10** CLT panel.

In structural systems, such as walls, floors and roofs, CLT panels serve as load-bearing elements. As such, in wall applications, the lumber used in the outer layers of a CLT panel is typically oriented vertically so its fibers run parallel to gravity loads, maximizing the wall's vertical load capacity. In floor and roof applications, the lumber used in the outer layers is oriented, and so its fibers are parallel to the direction of the span.

CLT's shear strength affords designers a host of new uses for wood. These include, wide prefabricated floor slabs, single-level walls and taller floor plate heights. As with other mass timber products, the CLT can be left exposed in building interiors, offering additional aesthetic attributes. In the tropics, however, the commercial manufacture of CLT is almost non-existent, although R&D efforts in this field are intensive in many Asian countries (Brandner et al., 2016).

Glue laminated timber (Glulam) comprises a number of wood laminates glued together (Fig. 11). The fibers in the laminates run parallel to the length of the piece. In straight glulam products, the laminates are 45 mm thick. For curved products, the thickness is less, generally 30 mm. The laminates are finger-jointed to produce long lengths and then bonded together to create the desired dimension. Glulam is a construction material that comes in a range

**Figure 11** Glulam.

of strength classes. The manufacturing standard for the Swedish market is strength class GL30. Some glulam beams are made by splitting glulam beams of class GL30, to create split timber beams, which then have a strength class of GL28.

> Glulam beams are made with laminates of a higher strength class on the bottom and top, where the maximum tensile and compressive stresses occur. The rest of the cross-section, where the stresses are lower, usually uses laminates of a lower strength class. The manufacturing method is called combined glulam, and is usually indicated with the letter 'C' after the strength class designation. This allows more efficient use of material compared with only using wood of the same strength class (FPL, 2001).

Glulam sets no limits on the potential for wood construction techniques. The glued laminates make glulam both strong and stiff. Glulam is one of the strongest construction materials in relation to its weight. This means that glulam beams can freely span large distances. Architects, structural engineers and users have great freedom to create their own shapes using glulam, whether for the structure of a house, the roof of a public building or a wooden bridge. Glulam is a construction material that optimizes the technical properties of the raw material of wood.

Glulam is more common in the tropics and has been used widely in building construction for some time now, although it is currently being explored for high-end wooden building construction (Ratnasingam et al., 2018). Nevertheless, it must be enunciated that due to the increasing effect of globalization, wood products manufacturing in the tropical region is not confined to tropical wood resources, but also incorporates a variety of imported temperate wood resources. Inevitably, the arising complex supply chain of the global wood products market for both non-structural and structural applications, does not preclude the tropical region, and is therefore a common feature of the global marketplace. Hence, suppliers and consumers alike must adopt to this complex supply chain and cope with its demand and ensuing trend, in order to ensure a sustainable and competitive wood products sector (Ratnasingam et al., 2018).

## 9 Biomass energy and co-generation

The world currently relies heavily on non-renewable fossil energy sources, such as coal, petroleum and natural gas. Nevertheless, long-term supply of fossil fuels remains uncertain beyond 2050, when energy demand is anticipated to pose problems throughout the world (Klass, 2011). As the availability of fossil fuels declines, the only renewable resource large enough to substitute for or replace fossil fuel, for the production of energy is biomass. In this context, many countries in the tropical region have embarked on policy and regulatory tools, to encourage more efficient and modern biomass systems through technological development and diffusion (Ratnasingam et al., 2018).

Industrial biomass includes energy systems generating electricity, heat, or liquid fuel from fuel-wood, agricultural crops, or manure. In 2015, biomass other than fuel-wood and charcoal has provided 8% of the global world energy (WEC, 2016). Current technologies for converting wood biomass into electricity and fuels include thermochemical and microbial processes such as combustion, gasification, liquefaction and fermentation (WEC, 2016). Biogas is most commonly produced using animal manure, mixed with water, stirred and warmed inside air-tight digesters that range in size from around 1 m³ for a small household unit, to as large as 2000 m³ for a commercial installation (WEC, 2016). The biogas can be burned directly for cooking and space heating or used as fuel in internal combustion engines to generate electricity. Examples of thermochemical processes include wood-fueled power plants, in which wood and woody wastes are combusted to produce steam that is passed through a turbine to produce electricity; the gasification of rice hulls by partial oxidation to yield fuel gas, which drives a gas turbine to generate electricity; and the refining of organic oils to produce diesel fuels. Another example is the alcoholic fermentation of corn to produce ethanol, which is then used in a variety of formulations in motor fuels (Klass, 2011). Soybeans and oil palms produce oil crops that can also be processed directly into biodiesel.

The combination of different biomass sources and conversion technologies can produce all the fuels and chemicals that are currently manufactured from fossil fuels. The major obstacle is the price competition from fossil fuels. Other renewable energy sources, including solar, wind, geothermal, hydropower, ocean resources and biofuels as well as hydrogen are being increasingly exploited for energy production to replace fossil fuels (IEA, 2016).

Perhaps, the main driving force for the push toward renewable energy sources is the increasing awareness of global climate change and its effects, as well as the urgent need to mitigate it. One of the most promising solutions is to use more wood biomass as the sustainable energy source, to replace the depleting, yet polluting fossil fuels.

## 10 Wood-based biofuels

The forest industry is a major user of biofuels derived from wood. Sawmills and pulp mills, both burn wood waste and mill residues, that cannot be converted into merchantable or commercial products. Co-generation of heat and electricity is common, and some mills even export electricity to the grid (FAO, 2010). In fact, using wood waste for fuel helps to reduce the dependency on fossil fuels.

However, harvesting wood primarily to produce wood-based biofuels, presents a different scenario. To determine whether harvesting wood for biofuels can reduce carbon dioxide emissions, other factors must be considered. First among these factors is the amount of carbon emissions associated with harvesting, transporting and using wood-based biofuels. Second, the long-term productivity of the land and its ability to replace the carbon stock lost to harvesting (Seifert, 2014) should also be considered. Finally, the biological changes resulting from continuous harvesting, such as change in stand age and soil fertility, may reduce productivity. Additionally, while the carbon emissions from harvesting wood can be offset with regrowth on the same land, the calculation of carbon foot-print should take into account the amount of carbon that could have been sequestered, if the trees were not harvested for biofuel production (Haberl et al., 2012).

It must therefore be emphasized that for commercial exploitation of wood residues to become a reality, 100% efficiency in wood conversion processes is a pre-requisite. In this case, residues or waste from one process would become the feedstock for the subsequent process. Under such circumstances, the aspiration of creating a circular economy based on the wood products sector may be within reach.

## 11 Conclusions

Although the tropical region covers a large tract of land in the world, the supply of tropical wood resources has been steadily on the decline over the years. In fact, the tropical region has the largest forest plantation area in the world, and it emphasizes its dependence on such man-made forests for its sustainable supply of wood resources. Apart from the forest plantation wood resources, increasingly large quantities of non-wood fiber resources, including bamboo, rattan and agricultural biomass are available in the tropical region, rendering it as the potential feedstock for traditional and new product development. In fact, it is apparent that the variety of fiber resources available in the tropics, offer a wide variety of option to manufacturers to replace traditional wood fibers with these new fiber resources, often leading to more competitive products, without sacrificing the quality. Apart from manufacturing panel products, chemicals, animal feed and other non-structural applications, there is a desire to incorporate engineered wood products into building construction to realize the green building objective, in line with the SDGs. Further, the use of wood for building construction and energy generation will have a significant impact on minimizing the carbon foot-print of these sectors, while inadvertently contributing towards mitigating the climate change effect. The tropical region of the world, with its large population, is embarking on an ambitious goal of shifting toward a circular economy through efficient use of wood and fiber resources, so as to ensure its compliance with the SDGs. In fact, such efforts backed by intensive R&D activities is beginning to yield positive results, and it may be anticipated that the future development of new wood products may indeed emerge from this region, driven by the variety of fiber resources available and the technological advancements to utilize them.

## 12 Where to look for further information

The Food and Agricultural Organization (FAO) of the United Nations publishes an annual State of the World's Forest report, which provides a comprehensive global overview of the forest resources and the current trends (www.fao.org/state-of-forests). The FAO also organizes the World Forestry Congress, which is the largest conference on the forestry sector.

The International Tropical Timber Organization (www.itto.int) is a United Nations linked organization that focuses primarily on tropical forests and forest products. It supports research at country-levels and published many useful reports on the subject.

The United Nations Economic Commission for Europe (UNECE) has a forestry and timber division (www.unece.org/forests) that also produces many reports on the subject. The Forest Products Annual Market Review is a worthwhile publication to look at for an overview of the trends.

The International Network for Bamboo and Rattan (www.inbar.int) is the leading global organization that provides information on these resources.

The International Society of Wood Science and Technology (www.swst.org) and the Forest Products Society (www.forestprod.org) are the two leading technical organizations in the world that disseminates information on wood and wood products through conferences, publications, trainings, symposia, etc.

The World Agroforestry Centre (www.worldagroforestry.org) and the Centre for International Forestry Research (www.cifor.org) also publishes scientific reports and disseminates relevant information on forests and biomass utilization trends.

The International Union of Forest Research Organizations (www.iufro.org) is the largest scientific cum academic network in the world, that interconnects the forests, science and people, for the overall development and sustainability of the forests and the environment.

## 13 References

Akil, H. M., Omar, M. F., Mazuki, A. A. M., Safiee, S., Ishak, Z. A. M. and Abu Bakar, A. 2011. Kenaf-fiber reinforced composites: a review. *Materials and Design* 32(8-9), 4107-21. doi:10.1016/j.matdes.2011.04.008.

Amar, K., Manjusri, M. and Lawrence, T. D. 2005. *Natural Fibers, Biopolymers and Biocomposites*. CRC Press Press, Taylor & Francis, London.

Ansell, M. P. (Ed.) 2015. *Wood Composites*. Woodhead Publishing, London, United Kingdom.

Bakri, M. K., Jayamani, E., Heng, S. K. and Kakar, A. 2018. Short review: potential production of Acacia wood and its biocomposites. *Materials Science Forum* 917, 37–41. doi:10.4028/www.scientific.net/MSF.917.37.

Bansal, A. K. and Zoolagud, S. S. 2002. Bamboo composites: material of the future. *Journal of Bamboo and Rattan* 1(2), 119–30. doi:10.1163/156915902760181595.

Brandner, R., Flatscher, G., Ringhofer, A., Schickhofer, G. and Thiel, A. 2016. Cross laminated timber (CLT): overview and development. *European Journal of Wood and Wood Products* 74(3), 331–51. doi:10.1007/s00107-015-0999-5.

Cao, Y., Shibata, S. and Fukumoto, I. 2006. Mechanical properties of biodegradable composites reinforced with bagasse fiber before and after alkali treatments. *Composites Part A: Applied Science and Manufacturing* 37(3), 423–9. doi:10.1016/j. compositesa.2005.05.045.

Csanàdy, E., Kovàcs, Z., Magoss, E. and Ratnasingam, J. 2019. *Optimum Design and Manufacture of Wood Products*. Springer Nature Publication, Switzerland AG, Geneva.

Davim, J. P. and Aguilera, A. (Eds.) 2017. *Wood Composite: Materials, Manufacturing and Engineering*. Walter de Gruyter GmbH & Co KG, Geneva, Switzerland.

Dransfield, S. and Widjaja, E. A. (Eds.) 1995. *Plant Resources of South-East Asia No 7: Bamboos*. Backhuys Publication, Leiden, Netherlands.

FAO. 2001. *Past Trends and Future Prospects for the Utilization of Wood for Energy*. Global Forest Products Outlook Study, Food and Agriculture Organization, Rome, Italy.

FAO. 2010. *Criteria and Indicators for Sustainable Woodfuels*. Food and Agriculture Organization, FAO Forestry Paper No. 160, Rome, Italy.

FAO. 2015. *Global Forest Resource Assessment – 2015*. Food and Agriculture Organization, Rome, Italy.

FAO. 2016. *Global Forest Products – Facts & Figures*. Food and Agricultural Organization, Rome, Italy.

FAO. 2019. *State of the World's Forests 2018*. Food and Agriculture Organization, Rome, Italy.

Feng, D., Caulfield, D. F. and Sanadi, A. R. 2001. Kenaf-fiber PP composites. *Polymer Composites* 22(4), 506–17. doi:10.1002/pc.10555.

FPL. 2001. *Wood Engineering Handbook* (2nd edn.). Forest Products Laboratory, Madison, WI.

Haberl, H., Sprinz, D., Bonazountas, M., Cocco, P., Desaubies, Y., Henze, M., Hertel, O., Johnson, R. K., Kastrup, U., Laconte, P., Lange, E., Novak, P., Paavola, J., Reenberg, A., van den Hove, S., Vermeire, T., Wadhams, P. and Searchinger, T. 2012. Correcting a fundamental error in greenhouse gas accounting related to bioenergy. *Energy Policy* 45, 18–23. doi:10.1016/j.enpol.2012.02.051.

Holbery, J. and Houston, D. 2006. Natural-fiber-reinforced polymer composites in automotive applications. *JOM* 58(11), 80–6. doi:10.1007/s11837-006-0234-2.

Hunter, I. R. and Junqi, W. 2002. *Bamboo Biomass*. INBAR Working Paper No. 36. International Network of Bamboo and Rattan, Beijing, China.

IEA. 2016. *Renewables in Global Energy Supply: An IEA Fact Sheet*. International Energy Agency, Paris, France.

IIED. 2017. *Towards a Sustainable Energy Path*. International Institute for Environment & Development, London, United Kingdom.

INBAR. 2017. *Socio-Economic Issues and Constraints in the Bamboo and Rattan Industry*. Working Paper No. 23. International Network for Bamboo and Rattan, Beijing, China.

ITTO. 2018. *ITTO's Sustainable Forestry Management (SFM) Action Plan*. International Tropical Timber Organization, Yokohama, Japan.

Klass, D. L. 2011. *Wood-Based Biomass for Energy Development in Sub-Saharan Africa – Issues and Challenges*. World Bank Publication, Washington.

Milner, H. R. 2009. Sustainability of engineered wood products in construction. In: Jamal Khatib (ed.), *Sustainability of Construction Materials*. Woodhead Publishing, London, United Kingdom, pp. 184-212.

Pande, H. 2013. Non-wood fiber and global fiber supply. *Unasylva* 49, 44-50.

Ratnasingam, J. and Awang, K. 2013. *A Review of Plantation Forestry in Tropical Asia – Potential and Challenges*. AIF Report No. 6. Bangkok, Thailand.

Ratnasingam, J. and Ioras, F. 2017. *A Critical Review of the Potential of Oil Palm Biomass as a Raw Material*. Report to the Ministry of Plantation Industries and Commodities. Putrajaya, Malaysia.

Ratnasingam, J. and Milton, T. 2011. *A Review of New Product Development from Wood and Agricultural Waste*. IFRG Report No. 14. Singapore.

Ratnasingam, J., Lim, C. L. and McNulty, T. 2012. *Rubberwood and Acacia Wood Resources – Prospects and Utilization in ASEAN*. IFRG Report No. 3. Singapore.

Ratnasingam, J., Ab Latib, H., Ng, W. C., Cellathurai, M., Chin, K. A., Senin, A. L. and Lim, C. L. 2018. Preference of using wood and wood products in the construction industry in peninsular Malaysia. *BioResources* 13, 5289-302.

Seifert, T. (Ed.) 2014. *Bioenergy from Wood – Sustainable Production in the Tropics*. Springer Nature Publication, Geneva, Switzerland.

Shmulsky, R. and Jones, P. D. (Eds.) 2019. *Forest Products and Wood Science: An Introduction* (7th edn.). John Wiley & Sons, Oregon.

Stokke, D. D., Wu, Q. and Han, G. 2014. *Introduction to Wood and Natural Fiber Composites*. John Wiley & Sons, West Sussex, United Kingdom.

Sunderland, T. C. H. and Dransfield, J. 2000. Species profiles in rattans. In: *Rattan: Current Research Issues and Prospects for Conservation and Sustainable Development*. Expert Consultation on Rattan Development, Non-Wood Forest Products No. 14. FAO, Rome, Italy.

WEC. 2016. *Global Annual Evaluation of Bioenergy Production*. World Energy Council, London, United Kingdom.

# Chapter 4

## Emerging technologies to develop new forest products

*Tatjana Stevanovic, Laval University, Canada*

## 1 Introduction

Forest trees occupy a central place among lignocellulosic plants, in terms of both abundance of wood they produce each year and also generating a variety of other biological tissues. These are rich sources of a multitude of natural products. The renewed interest in the development of new forest products resides in the need to replace exhaustible petroleum feedstock by renewables, while contributing to the reduction of greenhouse gas emissions and improving the carbon footprint of the final product. This goal is achieved by promoting the sustainable transformation of lignocellulosic biomass through chemical conversion. This approach can be visualized as a forest biorefinery, echoing a petrochemical refinery which converts petrol into chemicals and fuels. Established thermochemical processes, such as pyrolysis, carbonization and gasification, convert biomass into various fuels or commodities. In this chapter, we will focus on newer processes and new applications of wood polymers, notably uses of lignins.

The main objective of this chapter is the innovative valorization of lignin polymers and aromatic extractives present in lignified tissues of forest trees. All these developments are part of the concept of a forest biorefinery which can be defined as transformation of forest biomass into chemicals and fuels (energy). We will consider the three major biopolymers: cellulose, hemicelluloses and

http://dx.doi.org/10.19103/AS.2019.0057.25

lignins. A particular focus will be on aromatic polymers, notably lignins, as the wood constituents which are the richest carbon sources of all lignocellulosic polymers and the most important aromatic polymers on earth. These have been much less studied than cellulose.

## 2 Novel uses of cellulose

Cellulose is a fascinating biopolymer occupying a central place on earth. Its unique ordered structure, with a regular repetition of cellobiose fragments within extended linear chains in cellulose's native state, explains their affinity to assemble through low-energy intermolecular interactions into fibrillary structures of varying dimensions. This unique molecular and supramolecular structure of cellulose is the basis of its widespread occurrence in plants, in the form of a variety of fibrillary structures within which crystalline domains are intercalated by amorphous domains. This fibril structure contributes to the remarkable strength of cellulose-containing plant cell walls, including forest trees (Stevanovic, 2016). There is a growing interest in recovery of cellulose from a wide range of sources, from forest trees to agricultural residues (Szymańska-Chargot et al., 2017).

Traditionally, wood cellulose has mainly been obtained through a variety of chemical pulping processes. These apply harsh conditions to wood chips, including high temperatures and pressures, long hours of cooking in closed vessels and aggressive chemical systems, based mainly on sulphur-containing chemicals. Cellulose fibres are recovered in the form of solid cellulosic pulps, containing varying quantities of residual lignins and hemicelluloses. The technologies developed to transform pure cellulose into commercial esters and ethers require the application of bleaching procedures to minimize the residual lignin content.

New developments in cellulose pulp conversion into cellulosic nanomaterials represent the most important advances in novel cellulose applications. The development of nanotechnologies, which require pure cellulose samples, has marked a new era for cellulose applications in the form of cellulose nanoparticles in a variety of dimensions. The applications of these cellulose nanoparticles stem mainly from their extraordinary mechanical properties, due principally to the very important specific surfaces of these materials (Table 1). Some of these novel cellulosic nanoparticles are discussed below.

### 2.1 Microcrystalline cellulose

Microcrystalline cellulose (MCC) is defined as a material containing cellulose chains with an average degree of polymerization ranging between 200 and

**Table 1** Properties and uses of various cellulose nanoparticles

| Name | Dimensions | Applications |
|---|---|---|
| Microcrystalline cellulose (MCC) | Average DP 200-450 | Pharmaceutical as filler |
| Microfibrillated cellulose (MFC) | Diameters close to microns | Specialty paper products, composites and hydrogels |
| Nanofibrillated cellulose (NFC) | Diameter lower than 100 nm | Paper products, adult diapers |
| Cellulose nanocrystals (CNC) | Diameter of 10 nm | Personal care, food, specialty papers and so forth |

450. The MCC form colloidal solutions in water and are most used in the pharmaceutical industry as fillers or binders, for example for aspirin coatings (Rebouillat and Pla, 2013).

## 2.2 Microfibrillated cellulose nanoparticles

Microfibrillated cellulose (MFC) nanoparticles are produced by mechanical disintegration of cellulosic chemical pulps. This cellulosic material is used for composite materials, including paper products, based on its high mechanical resistance, which derives from its high aspect ratio and particular surface properties (Lepetit et al., 2017).

## 2.3 Nanofibrillated celluloses or cellulose nanofibrils

Nanofibrillated celluloses (NFC) are cellulosic nanoparticles produced by a combination of mechanical refining and chemical treatments which facilitate mechanical refining. Both of these treatments are usually applied to bleached chemical pulps, such as bleached Kraft pulp. These NFC are designed to have diameters lower than 100 nm, but it is difficult to overcome the inherent tendency of cellulose chains to assemble into larger fibrillary systems. An interesting approach to produce the NFC particles has been adopted by researchers at the Université de Troi-Rivières, Quebec, Canada (Mishra et al., 2011). The effect of 2,2,6,6-tetramethyl-1-piperidine oxoammonium (TEMPO) oxidation on bleached Kraft pulp was combined in that study with the mechanical action of ultrasounds to produce TEMPO-oxidized cellulose nanocrystals (CNCs).

The TEMPO technique itself is based on the selective oxidation of primary alcohol groups in glucose residues of cellulose chains into carboxylic acids (uronic acid groups, more precisely), using a TEMPO radical in the presence of sodium hypochlorite and sodium bromide (Saito and Isogai, 2005). Recently, NFC sheets produced by adding antibacterial and deodorizing functions to TEMPO-oxidized NFC were used to produce adult diapers by Nippon Paper Industries. The NFC cellulose, also known as CNF (from cellulose nano-fibres), is used in sheets with antibacterial and deodorizing functions in a range of skincare products. TEMPO-oxidized CNF is expected to be put into practical use for a wide variety of industrial applications, such as functional additives and nano-composite materials (Nippon Paper Group, 2017).

### 2.4 Cellulose nanocrystals

CNCs are produced by acid hydrolysis of bleached chemical pulps or of higher-sized cellulose particles, such as MCCs, MFCs or NFCs, to produce extended rod-like or whisker-shaped highly crystalline particles. CNCs are a novel class of nanomaterials with many interesting properties (Peng et al., 2011). The manufacturing principle of this cellulosic material is the acid hydrolysis of amorphous parts of bleached cellulosic pulp, leaving behind only crystalline ones. The diameter of CNC is about 10 nm, leading to specific surface area of 6000 $m^2$ $g^{-1}$. Various applications of CNC have been explored, for oilfield fluids and materials, for adhesives development (CNC enhances wet and dry strength of adhesives), improvement of paper quality, modifying the rheology of cements, in composite production for enhancement of mechanical properties, for paints and coatings (improvement of technological properties) as well as in personal care, health and food products (CelluForce, 2018). The main problem in CNC production resides in treatment of effluents, which are highly acidic, as well as in the very specific application of these materials, which require the identification of stable markets for these products, ahead of their larger-scale production.

## 3 Novel uses of hemicelluloses

Cellulose is found in wood cell walls with a number of heterogeneous polysaccharides, commonly designated as hemicelluloses. Hemicelluloses of temperate zone hardwoods are comprised mainly of acetylated glucuronoxylan and smaller quantities of glucomannan, while softwoods contain comparable quantities of arabino-glucuronoxylan and acetylated galactoglucomannans (Stevanovic, 2016). Due to their heterogeneous structure, and lower degrees of polymerization, wood hemicelluloses do not display any form of supramolecular organization in their native state. Hemicelluloses occur in the form of more or less branched polysaccharides, composed of pentose and hexose monosaccharides and their derivatives (mainly uronic acids). The application of hemicelluloses within a forest biorefinery stems mainly from the products of their conversions under pulping conditions, which are commonly found in the spent liquors (Stevanovic, 2016).

A historical example of pulping under acidic conditions is sulphite pulping. Under such conditions, xylans are hydrolysed to xylose monomers, which are further converted, still under acidic conditions, into furfural (Christopher,

2012). Monosaccharide xylose can be converted by hydrogenation into xylitol, an alditol. This hydrogenation process requires high pressure (50 atm) and temperatures (140°C) and use of Raney Nickel as catalyst. Xylitol can also be produced by microbial fermentation, using naturally occurring yeasts from genus *Candida*, such as *C. tropicalis* and *C. guilliermondii*. The fermentation yields are much higher, 60–95% when applying microbial fermentation xylan, with less-expensive purification procedures required (Prakasham et al., 2009).

Due to its anti-cariogenic and anti-plaque action, xylitol is widely used as a sweetener in chewing gums, pastilles and other products (Christopher, 2012). With 40% less nutritive energy, xylitol is used as a low-calorie alternative to table sugar (sucrose). Absorbed more slowly than sucrose, xylitol does not contribute to high blood sugar levels (hyperglycaemia) caused by insufficient insulin response. Xylitol's glycaemic index is approximately tenfold lower than that of sucrose. Xylitol consumption is also thought to have favourable effects on osteoporosis and bone health (Christopher, 2012).

There are other biotechnological conversions of simple sugars derived from hydrolysis of hemicelluloses from wood and other lignocellulosic sources. Citric acid can be obtained by fermentation with *Aspergillus niger*, on substrates mainly consisting of hexose sugars (including wood wastes). Citric acid is mainly used in the food and pharmaceutical industries. Lactic acid is another very important organic acid obtained by fermentation of both pentose (xylose) and hexose (glucose, mannose) sugars, with *Lactobacillus delbrueckii* or *A. oryzae*. It is used to acidulate, as a flavour enhancer, food preservative, as a feedstock for calcium stearoyl-2-lactylates (for baking), for ethyl lactate (a biodegradable solvent). It is also used for synthesis of polylactic acid (PLA) plastics which are used for packaging, consumer goods and in various biocomposites.

Succinic acid is yet another dicarboxylic organic acid which can be obtained by fermentation of glucose and xylose and is used to acidulate foods, to regulate pH, as flavouring and antimicrobial agent, as an ion chelator in electroplating applied to prevent metal corrosion, also used as a surfactant, detergent, foaming agent or for production of antibiotics and amino acids. Different substrates have been explored for succinic acid biotechnological production using *Actinobacillus succinogenes* for fermentation (de Barrosa et al., 2013). There are many other fermentation products from simple sugars, such as fumaric acid, itaconic acid, aspartic, malic acid and so forth, but a more detailed discussion about these would go beyond the scope of this chapter.

## 4 Novel uses of lignins

The lignins have long been neglected and regarded merely as by-products of pulping processes. However, lignins deserve to be treated as valuable biopolymers in their own right, the applications of which go far beyond simple fillers in composite materials. In addition to their adhesive properties, they are rich in carbon with the potential to be converted into various novel carbon-based products. One of the most studied applications of lignins is as adhesives in various panels and composite materials (Feldman, 2002). However, lignins are generally regarded as non-reactive polymers, which have poor performance in crosslinking reactions. Thus, commercial lignins are generally bad binders for wood panels in comparison to common resins such as phenol formaldehyde (Lewis et al., 1989). A number of lignin activation reactions have therefore been tried to enhance lignin reactivity as adhesive resins, among which a pre-demethylation (Alonso et al., 2005) or replacement of formaldehyde with glyoxal (Mansouri et al., 2007) or with activated lignin and isocyanate (Chauhan et al., 2014).

Our team has been investigating other potential applications of lignins from hardwoods and softwoods. We have explored Kraft lignin applications in high-density polyethylene composites (Schorr et al., 2014; Mariotti et al., 2014; Hu et al., 2015). Major challenges include the amount of lignin recovery from Kraft processes as well as impurities such as high ash (sulphur) and carbohydrate content (Lora and Glasser, 2002; Pan et al., 2005). This has highlighted the need to reconsider alternative processes such as organosolv pulping. Recently, we have been able to design a catalytic organosolv pulping process using an ethanol–water solvent system, with ferric chloride as a catalyst. The innovation of this patented process resides in the pre-extraction step, which is included to remove and recover phenolic extractives, thus protecting the delignification process (and the catalyst) from unwanted reactions, by impeding condensation reactions between lignin and phenolic extractives during pulping (Stevanovic and Koumba-Yoya, 2016).

The exploration of organosolv processes has also highlighted the potential for extracting lignins from bark (Koumba-Yoya and Stevanovic, 2017b). We have explored the adhesive performance of sugar maple wood and bark organosolv lignins in particle board fabrication. The sugar maple bark organosolv lignin used as a sole adhesive or in combination with isocyanate was more efficient than the corresponding formulations based on its wood counterpart, while the organosolv wood lignin combined with glyoxal performed better as adhesive than its bark counterpart. There is a huge potential in applying biopolymers from various tissues of forest trees including bark.

As an example of the range of applications of organosolv lignins, we have also explored the potential of ALCELL lignin, an organosolv lignin produced from pulping of mixed hardwoods with ethanol, in formulating hybrid coatings based on hydroxyapatite (HAP) for use on titanium bone replacements. HAP has similarities with human bone and is therefore a very suitable biocompatible material for bone implant coatings. However, HAP alone is a brittle material which is a major obstacle to its broader use. Research has focused on creating composites with biopolymers in order to improve the coating's adhesion performance. We have produced silver-doped hydroxyapatite and HAP coatings on titanium using electrophoretic deposition. Studies have shown that lignin addition improves the adhesion performance of the HAP ceramic coating on titanium (Eraković et al., 2012, 2013, 2014). The higher lignin concentrations were especially important for thermal stability of the coating, protecting the HAP lattice during high-temperature sintering.

Another promising potential application of lignins is in manufacture of carbon fibres. Carbon fibre is one of the toughest materials currently available (Franck et al., 2014). Carbon fibres are typically long and have remarkable mechanical properties, such as high-tensile strength and modulus of elasticity. Carefully designed carbonization processes are required to produce carbon fibres from polymers. One of the most commonly used polymers as a carbon fibre precursor is polyacrylonitrile (PAN). Pitch from petroleum feedstock is another common precursor of carbon fibre along with, at a lesser extent, regenerated cellulose fibres (e.g. rayon).

Unlike PAN or pitch, lignins are a renewable source of precursor polymers for carbon fibre production, with a carbon content typically higher than 60%. The majority of carbon-constituting lignins are $sp^2$-hybridized, which is the carbon hybridization state in carbon allotropes, graphene and graphite, giving them similar desirable properties. Different lignin activation procedures have been tried to modify lignins for electrospinning. These include acetylation (Zhang and Ogale, 2014), mixing with binder agents such as polyethylene oxide (Wang et al., 2013), polypropylene (Kadla and Kubo, 2004), polyvinyl alcohol (Ago et al., 2012), PLA (Thunga et al., 2014) and polyethylene terephthalate (Kubo and Kadla, 2005). Additional purification of lignin has also been explored as a means of improving the lignin conversion into carbon fibre, along with exploration of methods for stabilization of carbon fibre produced from kraft lignin (Norberg et al., 2013).

We have investigated the transformation of organosolv lignin-Fe by electro- and melt-spinning into lignin fibre, followed by thermal stabilization to transform these into carbon fibres (Koumba-Yoya and Stevanovic, 2016). The polymer properties of lignin-Fe were found to be superior to those of lignin-W or lignin-SA (lignins obtained from organosolv pulping without or with sulphuric acid as catalyst, respectively). The diameters of fibres produced from lignin-Fe ranged between 100 and 355 nm, and were more uniform than those reported in the literature for other lignins (Kubo and Kadla, 2005). We have also successfully performed melt-spinning experiments with pure organosolv lignin (lignin-Fe) without adding any other polymers or additives (Koumba-Yoya and Stevanovic, 2018). The diameters of fibres produced in these melt-spinning experiments performed with organosolv lignin (lignin-Fe) were determined to range between 20 and 35 μm, which are much larger dimensions though than those reported for fibres spun from PAN in similar experiments (5 and 10 μm (Liu et al. 2015a,b)). As lignin fibres need to be thermally treated for conversion into carbon fibres, we have evaluated the best parameters for their stabilization, consisting of thermal oxidation under air followed by carbonization under nitrogen, applying different conditions of heating rate.

Chemical analysis by XPS revealed that carbon fibres contained more carbon (96.20%), with C-C as major bond (85.42%), than lignin fibres which consisted of only 79.96% of carbon (result not presented here), but yet substantially more than original aspen wood organosolv lignin (Koumba-Yoya and Stevanovic, 2018). The increase of carbon content in carbon fibre as compared to lignin fibre is an indication of the efficiency of the thermal stabilization process and transformation of lignin fibre into carbon fibre.

Mechanical testing showed strength failure between 0.5 and 1.15 GPa, and elastic moduli evaluated between 70 and 114 GPa, which are higher values than reported so far in the literature for carbon fibres produced from any other lignin source examined for carbon fibre production (Baker et al., 2012; Baker and Rials, 2013). The maximum tensile strength determined for the carbon fibre produced from organosolv lignin (lignin-Fe) in our research is 1.24 GPa. We believe that such a high value can be explained mainly by the high purity of organosolv lignin used and its adequate thermal properties (high Tg). There is still work to be done to improve lignin-based carbon fibre performance, especially by further improvement of the thermal stabilization process.

Researchers have also investigated acetylation of softwood Kraft lignin to produce a carbon fibre with good tensile strength (Zhang and Ogale, 2014). The esterification of commercial Kraft lignin Indulin AT by butylation along with its mixing with PLA (at 25%) to improve the performance of fibre extrusion has also been reported (Thunga et al., 2014). Trials with Acetosolv lignin have also been reported by Utaki et al. (1995). Different procedures using lignin in combination with various binder polymers, such as polyethylene (Kubo and Kadla, 2005) or polyacrylonitrile (Liu et al., 2015a,b), have been explored to improve lignin transformation into carbon fibre by melt-spinning.

# 5 Polyphenols from bark

The dominant position of forest trees in the plant kingdom is mainly related to the strong lignification of their wood (and bark) tissues, which enables such exceptional growth, height and longevity of forest trees. Lignin biosynthesis occurs through a phenylpropanoid pathway, which provides, along with monolignols (cinnamic alcohol derivatives, lignin precursors), several other classes of polyphenols. The phenylpropanoid pathway is shared by several other phenolic metabolites which are therefore regularly found in wood, bark, roots and other tree parts and tissues, which have historically been well named by Bate-Smith as 'woody polyphenols' (1962). These comprise compounds with very specialized functions and structures. As they are mainly present in the form of free molecules, they are available for solvent extraction as wood (forest) extractives.

These extractable polyphenols include lignans and neolignans (structurally closely related to lignins), proanthocyanidins (condensed tannins), flavonoids and related compounds (coumarins), as well as stilbenes, the latter being important constituents of spruces and pines, for example. Lignins and extractable polyphenols also act as phytotoxins against fungi, bacteria and insects. The concentration of polyphenolic extractives is particularly high in bark as it is a protective tissue (Stevanovic et al., 2009). Many polyphenols found in forest trees have thus been determined to have important pharmacological activities in humans, such as anti-inflammatory, antioxidant, anticancer and have long history of such applications (e.g. salicylic acid from willows and poplar bark for aspirin). *In vitro* studies of black spruce bark extract, for example, have demonstrated its important anti-inflammatory and antioxidant properties and its role in boosting immune function (García-Pérez et al., 2013; Le Normand et al., 2014).

Our studies of extractives from black spruce (*Picea mariana*) bark have revealed a high concentration of resveratrol, along with a number of other phenolics and stilbenes and their glycosides, including astringin, isorhapontin and piceid (García-Pérez et al., 2012, 2013; Francezon et al., 2017; Francezon and Stevanovic, 2017). This offers the potential for developing products such as black spruce bark essential oil and aqueous bark extract for food, cosmetic and pharmaceutical applications.

Results of our ongoing study of extracts from maples: *Acer saccharum* M. (sugar maple) and *Acer rubrum* L. (red maple) bark and buds indicate that there is a potential of using these extracts, for example in the food industry (Meda et al., 2017; Geoffroy et al., 2017; Bhatta et al., 2018). The composition of water extracts from sugar and red maple bark are shown in Table 2. The food applications of maple sap and syrup, along with the use of maple sap in traditional medicines are the well-known examples of the long history of non-wood forest product applications as foods and health products. Maple syrup, a natural sweetener obtained by concentrating sap from sugar and red maple, has high nutritional value as it contains sugars, polyphenols, minerals, amino acids and vitamins. Some maple tree parts, and particularly so, its bark, have been used by Native People in the treatment of various diseases. There are also records about the use of maple bark infusions as drinks (Bhatta et al., 2018).

Catalytic organosolv pulping of sugar maple bark can transform bark into several valuable products. The process consists of pre-extraction with ethanol-water (50/50 v/v) of sugar maple bark followed (after removal of extractives)

**Table 2** Proximate and water-soluble sugar composition of water extracts from sugar and red maple bark

| Traits | Sugar maple bark (%) | Red maple bark (%) |
| --- | --- | --- |
| Moisture | 5.75 ± 0.09[a] | 5.30 ± 0.15[b] |
| Ash | 8.84 ± 0.05[a] | 3.40 ± 0.07[b] |
| Protein | 2.65 ± 0.13[a] | 1.50 ± 0.12[b] |
| Fat | 0.43 ± 0.12[a] | 0.36 ± 0.18[a] |
| Carbohydrates | 82.33[a] | 89.43[b] |
| Energy[A] | 343.80[a] | 366.97[b] |
| *Water-soluble sugars* | *(g/100 g DE)* | *(g/100 g DE)* |
| Complex sugars | 25.05 ± 1.38[a] | 18.85 ± 0.77[b] |
| Sucrose | 10.94 ± 0.11[a] | 5.58 ± 0.08[b] |
| Glucose | 5.28 ± 0.04[a] | 3.51 ± 0.06[b] |
| Fructose | 5.56 ± 0.03[a] | 4.50 ± 0.02[b] |
| Total sugar content | 46.83 ± 1.51[a] | 32.44 ± 0.92[b] |

Values represent means ($n$ = 3) ± SD; [a,b] different superscript letters in the same row are significantly different ($P < 0.05$) according to Holm-Sidak method; energy value expressed in kcal/100 g DE; DE: dry extract.
*Source*: adapted from Bhatta et al. (2018).

by pulping with the same solvent system, with ferric chloride as a catalyst. The conversion of a solid bark residue from sugar maple through an organosolv biorefinery can provide access to valuable biopolymers, cellulose and lignin, along with the constituents of residual liquor such as HMF which can be used for further organic synthesis (Koumba-Yoya and Stevanovic, 2016, 2017a,b; Yang et al., 2017). This example is yet another confirmation of an obvious need for the collaboration between the forest and chemical industries, which will allow access to new markets for forest products while improving the green aspect of chemical synthesis through introduction of intermediates derived from renewable resources.

## 6 Future trends and conclusion

As this chapter shows, forest biomass represents a real treasure of molecules with a wide range of potential uses. The collaboration between chemical and forest industries is indispensable, as it will bring benefits to both. The chemical industry is in need of green raw materials, processes and products and the work with biopolymers is a good starting point in this direction. The forest industry, on the other hand, has experienced important changes in the recent decades, evolving from an industry oriented towards commodity production into an industry seeking new value-added products, from construction through composites and non-wood forest products based on extractives to novel applications of wood polymers.

The focus of this chapter is on phenolic biopolymers lignins and on lower molar mass phenolic molecules belonging to forest extractives, some of which share the biosynthetic phenylpropanoid pathway with lignins and are regularly present in lignified tissues. Our studies on polyphenols from black spruce bark have revealed various stilbenes and their derivatives, including *trans*-resveratrol. As an illustration of polyphenolic extractives from hardwood bark, the hydrolysable tannins from red maple bark have been revealed as major polyphenols in red maple bark extracts applicable as natural health products. These results indicate the real potential of an integrated biorefinery approach that would allow for a simple water extraction of bark to be included in an existing wood mill.

A new catalytic organosolv biorefinery process can now provide access to a high-purity lignin, along with the production of solid cellulosic pulp, while phenolic extractives are recovered in the pre-extraction steps. We have presented the sugar maple bark biorefinery as an example of a specific application of bark organosolv lignin, for the adhesive formulation for wood panels. The carbohydrates and products of their transformation, furfural and HMF, were determined in residual liquor. There is also the possibility for catalytic conversion of HMF from residual liquor, derived from sugar maple bark organosolv pulping, into diformylfuran, as a potential intermediate furan derivative for organic synthesis of polymers (Yang et al., 2017).

Our results on transformation of organosolv lignin from aspen wood into carbon fibre are encouraging and there is real potential for transforming organosolv lignins into various carbon allotropes for high-value applications. In the presented studies, we have focused mainly on aromatic wood constituents. Future work on carbohydrates will follow in order to propose a complete conversion of wood biomass through a forest biorefinery. Our vision remains, however, that lignins are central biopolymers of forest trees, since considerable energy has been invested in their biosynthesis in plants. It seems therefore logical that lignins' conversion into high-value innovative products is just a question of intelligent design of technological procedures and fine-tuning of their parameters in order to reach the creation of high-value products. We are almost there!

## 7 Where to look for further information

We are continuing the research on the isolation and study of high-purity organosolv lignins from Canadian commercial woods in order to provide information on softwood lignins and their applications. The research on bark extracts will continue to attract attention and it will be important to make it possible to segregate barks from different tree species in order to provide access to barks of particular interest. The holistic approach to wood transformation is of interest, meaning that all wood constituents are adequately exploited. Smaller-sized mills will be designed so as to direct all production streams to well-identified customers. The solid cellulosic pulp will continue to be of interest for nanoparticles, while hemicelluloses will be available both from pre-treatment liquors and from the spent pulping liquors. Research centres such as Innventia AB, Sweden, or Oakridge Laboratory, Tennessee, are examples of research centres, along with numerous universities in Canada, France and elsewhere in the world.

# 8 References

Ago, M., Okajima, K., Jakes, J. E., Park, S. and Rojas, O. J. 2012. Lignin-based electrospun nanofibers reinforced with cellulose nanocrystals. *Biomacromolecules* 13(3), 918–26. doi:10.1021/bm201828g.

Alonso, M. V., Oliet, M., Rodríguez, F., García, J., Gilarranz, M. A. and Rodríguez, J. J. 2005. Modification of ammonium lignosulfonate by phenolation for use in phenolic resins. *Bioresour. Technol.* 96(9), 1013–8.

Baker, D. A. and Rials, T. G. 2013. Recent advances in low-cost carbon fiber manufacture from lignin. *J. Appl. Polym. Sci.* 130(2), 713–28. doi:10.1002/app.39273.

Baker, D. A., Gallego, N. C. and Baker, F. S. 2012. On the characterization and spinning of an organic-purified lignin toward the manufacture of low-cost carbon fiber. *J. Appl. Polym. Sci.* 124(1), 227–34. doi:10.1002/app.33596.

Bate-Smith, E. C. 1962. The phenolic constituents of plants and their taxonomic significance. I. Dicotyledons. *J. Linn. Soc. Lond. Bot.* 58, 95–173.

Bhatta, S., Ratti, C., Poubelle, P. and Stevanovic, T. 2018. Nutrients, antioxidant capacity and safety of hot water extract from sugar and red maple bark. *Plant Foods Hum. Nutr.* 73, 25–33. doi:10.1007/s11130-018-0656-3.

CelluForce. 2018. https://www.celluforce.com/en/applications/ (accessed on 25 October 2018).

Chauhan, M., Gupta, M., Singh, B., Singh, A. K. and Gupta, V. K. 2014. Effect of functionalized lignin on the properties of lignin-isocyanate prepolymer blends and composites. *Eur. Polym. J.* 52, 32–43. doi:10.1016/j.eurpolymj.2013.12.016.

Christopher, L. 2012. Adding value prior to pulping: bioproducts from hemicellulose. In: Okia, C. A. (Ed.), *Global Perspectives on Sustainable Forest Management*. InTech. ISBN: 978-953-51-0569-5. Available at: http://www.intechopen.com/book s/global-perspectives-on-sustainable-forestmanagement/adding-value-prior-to-pul ping-biofuels-and-bioproducts-from-hemicellulose.

de Barros, M., Freitas, S., Padilha, G. S. and Alegre, R. M. 2013. Biotechnological production of succinic acid by *Actinobacillus succinogenes* using different substrate. *Chem. Engin. Trans.* 32, 985–90.

Eraković, S., Veljović, Dj., Diouf, P. N., Stevanović, T., Mitrić, M., Janaćković, Dj., Matić, I. Z., Juranić, Z. D. and Mišković-Stanković, V. 2012. The effect of lignin on the structure and characteristics of composite coatings electrodeposited on titanium. *Prog. Org. Coat.* 75(4), 275–83. doi:10.1016/j.porgcoat.2012.07.005.

Eraković, S., Janković, A., Matić, I. X., Juranić, Z. D., Vukasinović-Sekulić, M., Stevanovic, T. and Miskovic-Stankovic, V. 2013. Investigation of silver impact on hydroxyapatite/

lignin coatings electrodeposited on titanium. *Mater. Chem. Phys.* 142(2-3), 521–30. doi:10.1016/j.

Eraković, S., Janković, A., Tsui, G. C. P., Tang, C. Y., Mišković-Stanković, V. and Stevanović, T. 2014. Novel bioactive antimicrobial lignin containing coatings on titanium obtained by electrophoretic deposition. *Int. J. Mol. Sci.* 15(7), 12294–322; doi:10.3390/ijms150712294.

Feldman, D. 2002. *Lignin and Its Polyblends—A Review.* Hu, T. Q. (Ed.). Springer, New York, NY, pp. 81–99.

Francezon, N. and Stevanovic, T. 2017. Integrated process for the production of natural extracts from black spruce bark. *Ind. Crops Prod.* 108, 348–54; doi:10.1016/j.indcrop.2017.06.052.

Francezon, N., Meda, N. R. and Stevanovic, T. 2017. Optimization of bioactive polyphenols extraction from *Picea mariana* bark. *Molecules* 22, 2118. doi:10.3390/molecules22122118.

Frank, E., Steudle, L. M., Ingildeev, D., Sporl, J. M. and Buchmeiser, M. R. 2014. Carbon fibers: precursor systems, processing, structure, and properties. *Angew. Chem. Int. Ed.* 53(21), 5262–98. doi:10.1002/anie.201306129.

García-Pérez, M.-E., Royer, M., Desjardins, Y., Pouliot, R. and Stevanovic, T. 2012. *Picea mariana* bark: a new source of *trans*-resveratrol and other bioactive polyphenols. *Food Chem.* 135, 1173–82.

García-Pérez, M. E., Allaeys, I., Rusu, D., Pouliot, R., Stevanovic Janezic, T. and Poubelle, P. E. 2013. *Picea mariana* polyphenolic extract inhibits phlogogenic mediators produced by TNF-α-activated psoriatic keratinocytes: impact on NF-κB pathway. *J. Ethnopharmacol.* 151(1), 265–78.

Geoffroy, T. R., Meda, N. R., Fortin, Y. and Stevanovic, T. 2017. Process optimisation for pilot-scale production of maple bark extracts, natural sources of antioxidants, phenolics and carbohydrates. Published online in Chemistry Papers.

Hu, L., Stevanovic, T. and Rodrigue, D. 2015. Comparative study of polyethylene composites containing industrial lignins. *Polym. Polym. Compos.* 23(6), 369–74.

Kadla, J. F. and Kubo, S. 2004. Lignin-based polymer blends: analysis of intermolecular interactions in lignin–synthetic polymer blends. *Compos. Part A Appl. Sci. Manuf.* 35(3), 395–400. doi:10.1016/j.compositesa.2003.09.019.

Koumba-Yoya, G. and Stevanovic, T. 2016. New biorefinery strategy for high purity lignin production. *ChemistrySelect* 1(20), 6562–70. doi:10.1002/slct.201601476.

Koumba-Yoya, G. and Stevanovic, T. 2017a. Study of organosolv lignins as adhesives in wood panel production. *Polymers* 9(2), 46. doi:10.3390/polym9020046.

Koumba-Yoya, G. and Stevanovic, T. 2017b. Transformation of sugar maple bark through organosolv biorefinery. *Catalysts* 7, 294. doi:10.3390/catal7100294.

Koumba-Yoya, G. and Stevanovic, T. 2018. Organosolv lignin for carbon fiber. In: Stevanovic, T. (Ed.), *Chemistry of Lignocellulosics: Current Trends.* Taylor & Francis Group, London, New York, pp. 214–30.

Kubo, S. and Kadla, J. F. 2005. Lignin-based carbon fibers: effect of synthetic polymer blending on fiber properties. *J. Polym. Environ.* 13(2), 97–105. doi:10.1007/s10924-005-2941-0.

Le Normand, M., Mélida, H., Holmbom, B., Michaelsen, T. E., Inngjerdingen, M., Bulone, V., Paulsen, B. S. and Ek, M. 2014. Hot-water extracts from the inner bark of Norway spruce with immunomodulating activities. *Carbohydr. Polym.* 101, 699–704. doi:10.1016/j.carbpol.2013.09.067.

Lepetit, A., Drolet, R., Tolnai, B., Montplaisir, D., Lucas, R. and Zerrouki, R. 2017. Microfibrillated cellulose with sizing for reinforcing composites with LDPE. *Cellulose* 24(10), 4303-12. doi:10.1007/s10570-017-1429-0.

Lewis, N., Lantzy, T. and Branhm, S. 1989. Lignin in adhesives: introduction and historical perspective. In: Hemingway, R. and Conner, A. (Eds), *Adhesives from Renewable Resources*. Oxford University Press, Oxford, UK, pp. 13-26.

Lin, J., Koda, K., Kubo, S., Yamada, T., Enoki, M. and Uraki, Y. 2014. Improvement of mechanical properties of softwood lignin-based carbon fibers. *J. Wood Chem. Technol.* 34(2), 111-21. doi:10.1080/02773813.2013.839707.

Liu, H. C., Chien, A. T., Newcom, B. A., Liu, Y. and Kumar, S. 2015a. Processing, structure, and properties of lignin- and CNT-incorporated polyacrylonitrile-based carbon fibers. *ACS Sustain. Chem. Eng.* 3(9), 1943-54. doi:10.1021/acssuschemeng.5b00562.

Liu, J., Willfor, S. and Xu, C. 2015b. A review of bioactive plant polysaccharides: biological activities, functionalization, and biomedical applications. *Bioact. Carbohydr. Diet. Fibre* 5(1), 31-61. doi:10.1016/j.bcdf.2014.12.001.

Lora, J. H. and Glasser, W. G. 2002. Recent industrial applications of lignin: a sustainable alternative to nonrenewable materials. *J. Polym. Environ.* 10(1/2), 39-48. doi:10.1023/A:1021070006895.

Mansouri, N. E., Pizzi, A. and Salvado, J. 2007. Lignin-based wood panel adhesives without formaldehyde. *Holz Roh Werkst.* 65(1), 65-70. doi:10.1007/s00107-006-0130-z.

Mariotti, N., Wang, X. M., Rodrigue, D. and Stevanovic, T. 2014. Combination of esterified Kraft lignin and MAPE as coupling agent for bark/HDPE composites. *J. Mat. Sci. Res.* 3(2), 8-22. doi:10.5539.

Meda, R. N., Suwal, S., Rott, M., Poubelle, P. E. and Stevanovic, T. 2017. Chemometrics-based approach to analysis of phenolic fingerprints of red and sugar maple bud extracts. *Austin Biochem* 2(1), 1009.

Mishra, S., Thirree, J., Manent, A. S., Chabot, B. and Daneault, C. 2011. Ultrasound-catalyzed TEMPO-mediated oxidation of native cellulose for production of nanocellulose: effect of process variables. *BioResources* 6(1), 143-212.

Nippon Paper Group. 2017. Press release. Available at: https://www.nipponpapergroup.com/english/news/year/2017/news170425003763.html.

Norberg, I., Nordström, Y., Drougge, R., Gellerstedt, G. and Sjöholm, E. 2013. A new method for stabilizing softwood kraft lignin fibers for carbon fiber production. *J. Appl. Polym. Sci.* 128(6), 3824-30. doi:10.1002/app.38588.

Pan, X., Arato, C., Gilkes, N., Gregg, D., Mabee, W., Pye, K., Xiao, Z., Zhang, X. and Saddler, J. 2005. Biorefining of softwoods using ethanol organosolv pulping: preliminary evaluation of process streams for manufacture of fuel-grade ethanol and co-products. *Biotechnol. Bioeng.* 90(4), 473-81. doi:10.1002/bit.20453.

Peng, B. L., Dhar, N., Liu, H. L. and Tam, K. C. 2011. Chemistry and applications of nanocrystalline cellulose and its derivatives: a nanotechnology perspective. *Can. J. Chem. Eng.* 89(5), 1191-206. http//doi:doi:10.1002/cjce.20554.

Prakasham, R. S., Sreenivas, R. R. and Hobbs, P. J. 2009. Current trends in biotechnological production of xylitol and future prospects. *Curr. Trends Biotechnol. Pharm.* 3, 8-36.

Rebouillat, S. and Pla, F. 2013. State of the art manufacturing and engineering of nanocellulose: a review of available data and industrial applications. *J. Biomater. Nanobiotechnol.* 4(2), 165-88. doi:10.4236/jbnb.2013.42022.

Saito, T. and Isogai, A. 2005. Ion-exchange behavior of carboxylate groups in fibrous cellulose oxidized by the TEMPO mediated system. *Carbohydr. Polym.* 61(2), 183-90. doi:10.1016/j.carbpol.2005.04.009.

Schorr, D., Diouf, P. N. and Stevanovic, T. 2014. Evaluation of industrial lignins for biocomposites production. *Ind. Crop. Prod.* 52, 65-73; doi:10.1016/j.indcrop.2013.10.014.

Stevanovic, T. 2016. Chemical composition and properties of wood. In: Belcacem, N. and Pizzi, A. (Eds), *Lignocellulosic Fibers and Wood Handbook. Renewable Materials for Today's Environment.* Wiley & Sons - Scrivener Publishers, pp. 49-106.

Stevanovic, T. and Koumba-Yoya, G. 2016. Organosolv process for the extraction of highly pure lignin and products comprising the same. WO 2016197233 A1, 15 December 2016.

Stevanovic, T., Diouf, P. N. and García-Pérez, M. E. 2009. Bioactive polyphenols from healthy diets and forest biomass. *Curr. Nutr. Food Sci.* 5(4), 264-95. doi:10.2174/157340109790218067.

Szymańska-Chargot, M., Chylińska, M., Gdula, K., Kozioł, A. and Zdunek, A. 2017. Isolation and characterization of cellulose from different fruit and vegetable pomaces. *Polymers* 9(12), 495. doi:10.3390/polym9100495.

Thunga, M., Chen, K., Grewell, D. and Kessler, M. R. 2014. Bio-renewable precursor fibers from lignin/polylactide blends for conversion to carbon fibers. *Carbon* 68, 159-66. doi:10.1016/j.carbon.2013.10.075.

Uraki, Y., Kubo, S., Nigo, N., Sano, Y. and Sasaya, T. 1995. Preparation of carbon fibers from organosolv lignin obtained by aqueous acetic acid pulping. *Holzforschung* 49(4), 343-50. doi:10.1515/hfsg.1995.49.4.343.

Wang, S. X., Yang, L., Stubbs, L. P., Li, X. and He, C. 2013. Lignin-derived fused electrospun carbon fibrous mats as high performance anode materials for lithium ion batteries. *ACS Appl. Mater. Interfaces* 5(23), 12275-82. doi:10.1021/am4043867.

Yang, Z., Qi, W., Su, R. and He, Z. 2017. Selective synthesis of 2,5-diformylfuran and 2,5-Furandicarboxylic acid from 5-hydroxymethylfurfural and fructose catalyzed by magnetically separable catalysts. *Energy Fuels* 31(1), 533-41. doi:10.1021/acs.energyfuels.6b02012.

Zhang, M. and Ogale, A. A. 2014. Carbon fibers from dry-spinning of acetylated softwood kraft lignin. *Carbon* 69, 626-9. doi:10.1016/j.carbon.2013.12.015.

# Chapter 5

## Agroforestry for the cultivation of nuts

*Michael A. Gold, University of Missouri, USA*

## 1 Introduction

Row crop agriculture, maize (*Zea mays*) and soybean (*Glycine max*) in particular, covers over 1.28 billion ha of land globally (FAO, 2017). Though extremely productive in terms of yield, these annual cropping systems are heavily dependent on external inputs such as energy, nutrients and pesticides, leading to a suite of ecological consequences (Wolz et al., 2017). Agroforestry includes an integrated set of designed land use practices that can provide economic production while simultaneously enhancing environmental services. As a form of multifunctional agriculture, agroforestry is unique in that it is tree based, adding strategic diversity at various scales in ways that can reduce threats and build resiliency under changing conditions.

Although not defined as such, agroforestry has been practiced worldwide for millennia throughout both the tropical and temperate regions (Smith, 1950; Zhaohua et al., 1991; Nair, 1993; Lelle and Gold, 1994; Lepofsky, 2009; Rossier and Lake, 2014). In the United States, temperate zone agroforestry as currently practiced is relatively recent (Gold and Hanover, 1987). At present, US

http://dx.doi.org/10.19103/AS.2018.0041.17

temperate zone agroforestry is defined as 'intensive land-use management that optimizes the benefits (physical, biological, ecological, economic, and social) from biophysical interactions created when trees and/or shrubs are deliberately combined with crops and/or livestock' (Gold and Garrett, 2009).

Agroforestry design is derived from the scientific discipline of agroecology, which draws upon ecological principles, functions and processes within perennial-based agricultural systems to create sustainable agricultural systems. Applying science to agricultural practices, agroecology and agroforestry focuses on both economic production and ecosystem enhancement (Krebs and Bach, 2018). Agroforestry design incorporates agroecological principles that have been under development for decades (Reijntjes et al., 1992; Vandermeer, 1995; Malézieux, 2012). Primary agroforestry and alley cropping design criteria include imitation of the structure and function of natural ecosystems with species that generate an economic yield to support the farmer (Krebs and Bach, 2018). Recommended principles for designing the cropping system based on this approach include selecting species for complementary functional traits, developing complex trophic levels and reproducing ecological succession (Malézieux, 2012).

Trait-based agroecology has emerged as one of the dominant paradigms in terrestrial ecology (Martin and Isaac, 2018). Studies that evaluate the complex linkages between functional traits, functional diversity and ecosystem functioning have recently begun to explore concepts surrounding 'multifunctionality', the idea that multiple ecosystem functions should be considered as a management goal. Multifunctionality has clear implications for trait-based agroecology and for agroforestry, where land managers manipulate on-farm diversity to enhance ecosystem services beyond yield alone (Finney and Kaye, 2017).

US agroforestry consists of five recognized practices: alley cropping, windbreaks, riparian and upland forest buffers, silvopasture and forest farming. Alley cropping, as one of the five recognized agroforestry practices, is defined as 'trees or shrubs planted in sets of single or multiple rows with agronomic crops, horticultural crops, or forages produced in the alleys between the trees that can also produce additional products' (USDA-FS, 2017). This chapter will primarily focus on the opportunities and challenges associated with alley cropping practices involving overstorey nut crops as one element of the solution to address global needs for economically viable food production while enhancing environmental services (Wolz and DeLucia, 2018).

## 2 Nut-based agroforestry systems

Alley cropping can be viewed as recreating the structure of a savanna in an agroforestry system with multiple canopy layers consisting of a design that contains an overstorey of nut trees, mid-layer of fruiting small trees and shrubs, and a groundcover of annual crops or perennial grass groundcover. In nut-based agroforestry systems, crops grown in the alleys between tree rows provide annual income while the longer-term nut crop matures. When speciality crops such as herbs, fruits, vegetables, nursery stock or flowers are grown in the alleys, the microclimate created by the trees enhances the economic production of these sensitive high-value crops in stressed environments (Garrett et al., 2009).

It will typically take 8–15 years from the time of establishment of a cultivar-based nut orchard until reliable commercial yields, that is, there are enough nuts to warrant mechanical harvesting. This long time lag is a serious issue when it comes to cash flow and profit generation. For this reason nut orchard growers are turning to alley cropping, that is, establishing intercrops between tree rows until the overstorey trees generate positive cash flow. Many agronomic crops can be successfully intercropped between young chestnut, pecan or black walnut trees including corn, soybeans, winter wheat and milo. Once commercial nut tree yields occur, a shift to hay crops (i.e. clover/grass mixes) can be used to facilitate mechanical nut harvest while continuing to generate income from hay sales.

High-value nut or acorn-producing hardwoods such as walnut (*Juglans* spp.) (Grant et al., 2006; Scott and Sullivan, 2007; Reid et al., 2009; Coggeshall, 2011b; Geyer and Fick, 2014; Cardinael et al., 2015), chestnut (*Castanea* spp.) (Hunt et al., 2012; Wilson et al., 2018), pecan (*Carya illinoensis* (Wangenh.) K. Koch.) (Bugg et al., 1991; Kremer and Kussman, 2011; Fletcher et al., 2012; Van Sambeek and Reid, 2017), hazelnut (*Corylus* spp.) (Molnar et al., 2005) and oak (*Quercus* spp.) (Eichhorn et al., 2006) are the favoured species in alley cropping practices (Molnar et al., 2013; Wilson et al., 2018).

Primary economic benefits and uses for nut-based agroforestry systems include the sequential and integrated production of annual and higher value long-term crops and enhancement of microclimate conditions to improve crop or forage quality and quantity. Environmental benefits include reductions of surface water run-off and erosion; decreasing offsite movement of nutrients or chemicals improvement of soil quality by increasing utilization and cycling of nutrients; and enhancement of habitat for wildlife and beneficial insects (USDA-FS, 2017).

A key advantage of food-producing agroforestry systems is that they are more resilient to climate stressors than are annual cropping systems. Compared to temperate zone annual crop monoculture production practices, integrated tree and crop practices, that is, alley cropping, can increase land use efficiency through improved overall yields compared to separate production of trees and crops in monocultures (Wolz et al., 2017; USDA-FS, 2017). Agroforestry systems can diversify farm portfolios, spread risk and increase both the economic and environmental resilience of farms (USDA-FS, 2017; Krebs and Bach, 2018).

In addition, alley cropping can increase ecosystem resiliency by buffering against weather extremes, diversifying income as a hedge against financial risk, increasing biodiversity, reducing soil erosion and improving nutrient- and water-use efficiency (USDA-FS, 2017; Krebs and Bach, 2018).

The production of nuts and other speciality crops grown in agroforestry systems is increasing in direct response to increasing demand for locally sourced foods (Low and Vogel, 2011; Johnson, 2014; Jose et al., 2018; USDA-ERS, 2015). Fruit, nut and vegetable farms are eight times more likely to sell locally than are other farms (Low and Vogel, 2011). According to statistics collected by the USDA Agricultural Marketing Service, the number of farmers' markets has increased from 1755 to more than 8687, reflecting a 5-fold increase since 1994 (USDA-ERS, 2018). Fruit, nut and vegetable crops produced within agroforestry systems can become a major contributor to increasingly popular direct-to-consumer food systems.

## 3 Key challenges facing nut-based agroforestry systems

In spite of the strong economic and ecological case justifying the increased adoption of agroforestry and nut tree alley cropping (Jose, 2009; Rigueiro-Rodriguez et al., 2009; Udawatta and Jose, 2012), these systems have not been widely adopted in many regions of the United States where annual monoculture crops dominate the landscape (Valdivia et al., 2012; Moreno et al., 2018). A number of key issues and challenges must be addressed for the promise to become a reality including:

- industry grower support and networks;
- long-term research to select and improve nut tree cultivars across geographic locations;
- detailed production and management information about alley cropping configurations, including efficient mechanical harvest options;
- lack of solid market and financial information;
- policy support.

A knowledge gap exists in terms of what is known regarding the scientific underpinnings and associated practical implementation information for both some of the individual nut tree species in the eastern United States (e.g. chestnut, eastern black walnut, hazelnut) and orchard speciality crops as well as knowledge about design and management of integrated nut tree-based alley cropping systems (Kahn et al., 2011).

There is a need for guidance from extension services in land grant universities. This knowledge network also needs support from other sectors, for example, equipment dealers, private consultants, government agencies,

NGOs, private sector firms that will invest in nut tree alley cropping ventures, as well as farmer-to-farmer networks.

The appeal and viability of agroforestry is substantially undervalued when the productive potential of nut tree crops is not fully explored and exploited. The integration of nut tree species in alley cropping systems, especially those nut trees that have been genetically selected for precocity and high yields across locations, offer the opportunity to improve agroforestry systems' revenue potential and rate of return, enhancing their appeal for prospective adopters (Lovell et al., 2018). There is a need for continued genetic improvement of specific nut tree species and nut tree and companion crop interaction studies.

Production challenges include what to intercrop, how to manage tree and crop competition above and below ground (spatially) and through time (temporally) as the tree canopies develop and eventually dominate the site, how to manage the nut tree crop differently from the annual crops (irrigation, fertility, weed control, pest and disease controls) and how to efficiently mechanically harvest the tree crops within an alley cropping configuration.

To reduce financial risk and enhance adoption there is a need for accurate financial decision support tools and market and consumer information to provide solid growth, yield and price information to use with lenders for bank loans. Finally, there is a need for policy support, including the availability of crop insurance against losses and grants, incentive payments and low interest loans to help support innovative agroforestry production systems.

## 4 Genetic improvement of nut trees

Long-term research supporting the integration of nut trees into agroforestry systems is underdeveloped. Recommended actions include the development of multiple research trial locations using Land Grant research centres as sites for stable, long-term comparison trials (Lovell et al., 2018) and an increase in USDA funding for public plant breeding (this type of funding existed up to the 1980s) (Shelton and Tracy, 2017). The potential for nut trees to produce a wide range of commercial products and materials has grown recently due to germplasm collection and nascent genetic improvement efforts of edible nut species that have been historically underutilized in the Midwest United States. In this region, none of the important tree nut species, including chestnut, black walnut, northern pecan, hazelnut and oak, have achieved their full potential for increased productivity, disease resistance and broader environmental adaptability (Lovell et al., 2018).

To date, limited numbers of improved selections of chestnut, black walnut, northern pecan, hazelnut and oak germplasm have or soon will be transitioned into replicated performance trials in the Midwest region. This is in response to the need to assess the adaptability of cultivars to different regions (Smith, 1950;

Molnar et al., 2013). Initial selection and breeding work has been formative in developing first-generation nut trees, (Reid et al., 2004; Hunt et al., 2005; Molnar et al., 2005; Coggeshall, 2011a; Anagnostakis, 2012; Molnar and Capik, 2012; Wood et al., 2016) allowing a broader collection of commercially suited nut species to be considered when conceptualizing alley cropping plant assemblages. Improved nut tree cultivars, including further improvement of rootstocks (Grauke and Thompson, 2003), will boost the overall performance of the system.

## 5 Management of temporal and spatial tree and crop interactions

Alley cropping is designed to provide an increase in total productivity (i.e. considering both crop and tree production) (Muschler, 2015). Complementary functional traits of alley cropping tree/crop components are used to ensure both improved production and resilience. Different use of resources by different species, occupying different ecological niches, forms the basis of this concept of complementarity. Complementary phenology, coupled with strategic management activities, both temporal and spatial, are effective ways to optimize positive interactions between trees and crops for key resources such as light, water and nutrients while minimizing competition (Malézieux, 2012; Lovell et al., 2018; Nicholls et al., 2016; Wells et al., 2018; Wilson et al., 2018).

Agroforestry field experiments are designed to quantify competitive (negative) and facilitative (positive) relationship interactions between trees and crops and/or animals (Jose et al., 2004). Decisions about the experimental design include specific species/varieties and arrangement of species within treatments, necessary size of plots, number of replications to provide statistical significance and arrangement of plots in the landscape (Malézieux, 2012). At the farm level, the practical implementation of these principles differs and designs are specific to each specific location and situation.

Agroforestry systems are dynamic and interactions among system components (i.e. trees and crops), temporal changes, can be expected over the life of the system and will change over time. Although initial complementarity in sharing light, water and nutrient resources can be expected, over time, there will be an overlap between the needs and physical structures of overstorey nut trees and associated annual crop species. In agroforestry and alley cropping systems, this is to be expected. Complementarity can initially result in overyielding, but competition intensifies with time. Strictly in terms of annual crop yield, alley cropping usually results in a decrease in crop yield over time compared to the pure crop because as the trees mature, above and below ground, competition for light, water and nutrients increases between the crop and tree. Thus, agroforestry systems experience a complex series of inter- and intra-specific interactions guided by modification and utilization of light, water and nutrients (Gillespie et al., 2000; Jose et al., 2004, 2009; Thevathasan et al., 2004; Lacombe et al., 2009; Fletcher et al., 2012; Bainard et al., 2013; Cardinael et al., 2015; Inurreta-Aguirre et al., 2018a,b).

Long-term negative yield effects of temporal and spatial tree and crop interactions can be managed not only through complementary seasonal phenology between trees and crops, but also mitigated through annual site manipulation. To prevent tree and crop interaction below ground, when nut trees are planted to establish an alley cropping system, it is feasible to impose annual deep 'ripping' down to 12–18″ (put in cm) (1–2 m) from the base of the trees.

A research trial of two Midwestern US alley cropping systems showed no yield decrease in maize (a C4 species) in response to shading (Gillespie et al., 2000). In general, the eastern-most row of maize in a black walnut (*Juglans nigra*) alley cropping system received 11% lower photosynthetically active radiation (PAR) than the middle row. Irrespective of shading, no apparent yield reduction was observed when below-ground competition for nutrients and water was eliminated through trenching and polyethylene barriers. In contrast to black walnut, shading was greater in a red oak (*Quercus rubra*) alley cropping system due to higher canopy leaf area and temporally longer seasonal duration in full leaf during which a 41% reduction in PAR was observed for the eastern row. Similarly, western rows were receiving 17% and 41% lower PAR in comparison to the middle rows in the black walnut Jose et al. (2004).

Double tree-row alley cropping systems have also been designed for pecan to create a viable compromise between trees and crops both in terms of crop management and yield and long-term tree management and yield (Van Sambeek and Reid, 2017).

Lovell et al. (2018) have described long-term agroforestry systems research at two long-term experimental farms, Domaine de Restinclières, located near Montpellier, France, and the University of Missouri Horticulture and Agroforestry Research Center (HARC), located in New Franklin, Missouri, USA. The experiment at Domaine de Restinclières provided important spatial and temporal findings. Tree row orientation resulted in differences in radiation to the alley crop with north–south tree rows providing more homogeneous PAR to the grains. Using a winter wheat alley crop demonstrated temporal and phenological complementarity as the trees and crops did not directly compete for light during much of the year. Winter wheat grows from late fall into early summer, whereas walnut trees bud out in spring/early summer and nut harvest occurs in the fall prior to sowing the winter wheat. Over 20 years, on average, crop yields on agroforestry sites were only reduced 2% compared with monoculture controls. All alleys at Restinclières were managed with commercial-scale farm equipment both for ease of management and directly representative of large-scale application.

Results from long-term chestnut, black walnut and pecan germplasm trials at HARC in Missouri also offer specific information on species selection and management. For example, results of specific trials on chestnut pollination, for both variety compatibility and spacing/density requirements, provide guidance for future alley cropping designs. Higher yields and nut sizes in orchards under trickle irrigation have demonstrated the need for supplemental moisture during periods of summer drought. Research findings from both Restinclières and HARC have informed the development of the recently established Agroforestry for Food field research trial at the University of Illinois designed to study multifunctional woody polyculture systems. The Agroforestry for Food field trial seeks to address several key themes for the future of agroforestry research: food security, climate change, multifunctionality and practical applications.

Augmenting traditional alley cropping through multifunctional woody polyculture systems that include tree crops for food and fodder enhances the potential of alley cropping as a transformative solution to the problem of agriculture across the temperate zone. These systems provide many economic and ecological advantages over conventional row crop agriculture. Key economic drivers of these multifunctional woody polyculture systems of alley cropping include overyielding, utilization of crop analogues compatible with existing staple crops and resilience via crop diversification. Key ecological benefits include enhanced carbon sequestration, soil and nutrient stabilization, biodiversity and resilience to ecological pressures (Wolz et al., 2017).

Effective integration of multifunctional woody polycultures and tree crops in temperate alley cropping requires strategic implementation on marginal lands, an emphasis on highly productive tree crops, practical and optimized multispecies designs and complementary crop combinations for early productivity and management efficiency (Wolz et al., 2017).

## 6 Orchard design and management

One of the first problems that growers face in planning a new orchard/alley cropping system is deciding on the spacing for the trees – a decision with long-term implications. Growers planning to farm between tree rows need to think about the width of their alleyways. Between-row spacing is primarily determined by machinery width, mature size of nut tree species, with secondary considerations for sunlight requirements of the alley crop. With closer spacing, the intercrop grown in the alley will encounter light competition at an earlier age, requiring a more rapid change to shade-tolerant intercrops (Walter et al., 2015; Wilson et al., 2018). Wider spacing makes intercropping easier but delays the fullest realization of economically viable nut crop yields. Spacings ranging from ~20' by 20' to ~60' by 60' are recommended depending on both the choice of nut tree species (Coggeshall, 2011b; Hunt et al., 2012), the width of the farm equipment needed for planting and combining the intercrop between the tree rows and the willingness of the landowner to thin out overstorey trees within or between rows over time.

Alley cropping, as with other forms of agroforestry, requires both intensive technical management skill and marketing knowledge (Table 1). Among the many practical considerations, alley cropping with nut trees will require countermeasures against deer and rabbits in the form of either individual tree cages and/or establishment of permanent 8-10' high fencing around the entire perimeter of the orchard in the long term. Within tree rows, weed control is essential during the establishment years either through the use of herbicides or mulching.

In the Midwest United States, where extended periods without summer rainfall are increasingly common due to more intense climate fluctuations, including an irrigation or fertigation system in the overall design and management is essential for optimizing long-term nut tree productivity. This is true for Chinese chestnut (*Castanea mollissima*) in Missouri where adequate supply of soil moisture greatly improves overall yield and nut size, and may (untested) also improve yield of other tree nut species of interest.

## 7 Pest management in nut tree alley cropping

There is evidence that biodiverse plant communities, characteristic of natural ecosystems, can be used in agroforestry systems to improve pest management (Bugg et al., 1991; Shapiro-Ilan et al., 2012; Stamps et al., 2009). Stamps et al. (2009) investigated insect pest dynamics, crop yields and small farm economics

**Table 1** Considerations for the landowner prior to establishment of nut tree alley cropping

Alley cropping, as with other forms of agroforestry, requires both intensive technical management skill and marketing knowledge.

The following limitations should be considered:

- Removes land from annual crop production and may not provide a financial return from the trees for several years
- Requires a more intensive management system including specialized equipment for the tree management and additional managerial skills and training to manage multiple crops on a given site
- Requires a marketing infrastructure for forest products that may not be presently established
- Trees may be an obstacle to crop cultivation if not carefully planned and managed
- Trees compete with companion crops for light, moisture and nutrients
- Companion crops may compete with trees for moisture and nutrients
- Herbicide drift from crops may damage trees

*Source*: adapted from Walter et al. (2015).

in an alley cropping practice of alfalfa and black walnut compared to conventionally grown alfalfa. Specifically they examined alfalfa weevil, *Hypera postica* (Gyllenhaal), and its natural enemies in forage grown in an eastern black walnut (*Juglans nigra* L.) alley cropped system and in a traditional monoculture system. The presence of walnut trees increased natural enemy numbers and significantly increased parasitism of alfalfa weevil. Their research demonstrated that alfalfa intercropped with walnut supported significantly more parasitic hymenoptera and/or predators than did traditionally grown alfalfa weevil. Alfalfa weevil mortality from parasitoids and/or fungal infection was generally higher for weevil larvae from alley cropped alfalfa compared to weevils from monocropped alfalfa.

Pecan (*Carya illinoinensis* (Wangenheim) K. Koch) is an economically important nut crop in North America. The pecan weevil, *Curculio caryae* (Horn), is a key pecan pest throughout the southeastern United States stretching into Texas, Oklahoma, Kansas and Missouri. Control recommendations consist mainly of above-ground applications of chemical insecticides (e.g. carbaryl, pyrethroids) to suppress adults. Application is recommended every 7–10 days during peak emergence (generally up to at least a 6-week period). Fungi such as *Beauveria bassiana* (Balsamo) constitute potential alternatives for *C. caryae* control. Shapiro-Ilan et al. (2012) found that persistence and efficacy of endemic *B. bassiana* can be enhanced by the addition of a clover cover crop within the pecan orchard.

## 8 Financial decision support tools

There is a need to provide solid growth, yield and price information in the form of financial decision support tools and market and consumer information to reduce financial risk for producers and, in turn, to use this detailed information when approaching lenders for bank loans. The USDA National Agriculture Census, providing a wealth of data on all commodity and most speciality crops, added chestnut to the survey for the first time in 2007 and at present no data are collected on eastern black walnut as a nut crop (in contrast to the well-established 'California walnut' industry which is included in the National Agriculture Census). In order to provide some baseline information, Gold et al. (2006) and Aguilar et al. (2009, 2010) have conducted a series of market and consumer studies on the emergent US chestnut industry to supply reliable market trend and consumer preference information. Miller et al. (2016) interviewed owners, managers, consultants, board members and affiliates of five Midwestern businesses engaged in regional nut processing to provide industry leaders with information gleaned from the experience of established Midwestern nut processors and marketers.

The application of agroforestry can be greatly enhanced through the use of financial decision support tools to help integrate diverse sets of information. A variety of decision support tools addressing biophysical, economic and social considerations are available for applying agroforestry (USDA-FS, 2017). Godsey and others have created financial decision support tools for chestnut and elderberry which are designed to assist with establishment and management decisions. The tools also enable landowners to develop enterprise budgets and combine these budgets into annual cash flow plans for evaluation (http://www.centerforagroforestry.org/profit/#budget).

## 9 Policy support

Additional policy support for nut tree alley cropping, including the availability of crop insurance against losses and grants, incentive payments and low interest loans, is needed to help support innovative agroforestry production systems. Valdivia et al. (2012) examined the barriers preventing landowner adoption of agroforestry practices. Two overarching barriers were identified; one related to transaction costs in establishment of trees, and a second related to profitability concerns. Transaction costs exist due to the weakness of institutions that facilitate access to information, in this case regarding agroforestry practices and policies that can support implementation. Access to information and cost-share programmes would reduce the barriers to adoption of practices that incorporate trees on the landscape, by reducing transaction costs of establishing the practice. Incentives need to include reduction of costs and increased profitability to be appealing to those more engaged in farming.

Cartwright et al. (2017) describe an innovative, pilot-scale cost-share funding opportunity offered through the USDA NRCS Environmental Quality Incentives Program in Missouri. This special funding pool targets landowners interested in establishing agroforestry and/or woody crops on their property to encourage agroforestry adoption. In addition, the USDA Sustainable Agriculture Research and Education Program offers a variety of competitive grant opportunities to support both research and landowners who wish to engage in sustainable agriculture including support for agroforestry, speciality crops and nut tree alley cropping (https://www.sare.org/Grants).

## 10 Case studies

Pantera et al. (2018) describe the components, structure, ecosystem services and economic value of ten case studies where high value trees (i.e. fruits, nuts, timber) are integrated with understorey crops and/or livestock in Europe. Examples include agroforestry systems with olive, walnut, chestnut, apple and pear. Overall, the value derived from overstorey trees in these systems (olives, apples, nuts, chestnuts) were the primary focus of the farmers with grazing or intercropping of secondary importance. In addition, chestnut woodlands are identified by local stakeholders as one providing good habitat for the commercial production of edible mushroom. Trees grown on a regular grid pattern permitted use of machinery that enabled high outputs for commercial production, while systems with low inputs and levels of mechanization, such as chestnuts in Spain, also delivered high outputs.

This section focuses on two case studies of nut tree alley cropping, starting with Shepherd Farm (Flaim, 2005). Shepherd Farms is in north central Missouri on the Chariton River. The family first began farming there in the late 1960s, growing corn, beans and wheat on 1900 acres. In the early 1980s the farm shifted production to alternative crops. The philosophy behind the switch to a sustainable farming venture that included alley cropping with nut trees is embodied in the following quote from Dan Shepherd's father 'We stand to make a little money doing what others are already doing, or we can make a lot of money doing things others won't'.

According to the Shepherd's marketing of conventional commodities and livestock production means that someone else sets the price. Shepherd Farms switched production to include 270 acres of improved grafted cultivars of alley cropped pecan, pastured buffalo and eastern gamagrass seed production – three unique enterprises for which they could set the price and sell directly to the consumer via their on-farm store. For the first 20 years, positive cash flow in pecan orchards was obtained via alley cropping. They grew wheat and soybeans for small returns that basically paid for their expenses. As the trees matured the Shepherds seeded the alleyways with blue grass. The grass offered a great feed for the Shepherds' buffalo while creating an excellent surface for harvesting over 200 000 pounds of pecans.

The second case study is Hopeton Farms which includes 2100 acres of almonds and 65 acres of organically grown walnuts, located in California's Central Valley. The farm started in the mid-1980s as a partnership of three non-farming landowner families. The owners' interest in creating a more environmentally sound enterprise, involved a 28-acre block of the farm's almond trees in a 'BIOS' (Biologically Integrated Orchard Systems) pilot project initiated by the California Alliance with Family Farmers (CAFF), a California nonprofit organization to develop more sustainable orchard systems. Initial soil tests revealed nutrient-depleted soils and as a result they applied composted dairy manure over the whole farm (Wechsler, 2005).

California almond orchards typically have little-to-no plant cover on the soil, and what is there is mowed very close for ease of harvest. The Hopeton owners decided to try cover crops for soil improvement. In the first year of the project, they experimented with three different cover crop mixes to find the best combination. A low-growing mix, made up of annual clovers and vetches with a small amount of low-growing grasses, worked best. They settled on this, plus an insectary mix every tenth row. Following the success of the original test plot of 28 acres, the owners have added 300–400 acres of almonds under cover crops each year. In addition, instead of removing and burning tree prunings, they now chip up the prunings and put the wood back into the soil. Hopeton

Farms saved US$375 000 in pesticide and fertilizer input costs by cutting back on pesticides and fertilizers in the first year (Wechsler, 2005).

With the dedicated use of compost and ground covers and reduced use of soil chemicals, soil organic matter levels improved, with the return of earthworms throughout the orchard as evidence of improved soil health. With a more diverse habitat, raptors, owls and hawks also returned to the farm (Wechsler, 2005).

## 11 Conclusion and future trends

Agroforestry and nut tree alley cropping provide transformative solutions beyond monoculture grain production as the only primary food product. Agroforestry systems offer options for mitigation by reducing GHG emissions compared with other agroecosystems, and for adaptation through greater resilience under variable environmental conditions. Multifunctionality seeks to bring ecological and cultural functions into a production system, offering a new alternative for growers in temperate zones. Practical applications are needed to move agroforestry and alley cropping with nut trees beyond basic sciences, to study practical options for growers (Mori et al., 2017; Lovell et al., 2018).

In spite of encouraging initiatives related to soil health, cover crops, precision agriculture and genetic modification of annual crops to increase yield, transformative solutions that address the fundamental issues associated with vast monocultures of annual crops will be necessary for robust and resilient agricultural land use, especially in the face of climate change (Wolz et al., 2017). Agroforestry, including nut tree alley cropping, is one of the transformative approaches in need of further study in the coming decades. If nut tree alley cropping practices are to become one of the successful transformative solutions, they must become proven practices that are ecologically sustainable, economically viable and culturally acceptable. Ecological sustainability requires robust functioning of regulating and supporting ecosystem services alongside the provisioning services at the core of agriculture.

Luedeling et al. (2011a) projected climate change effects upon agroforestry systems that include temperate fruit and nut trees. Concerns exist about early bud break followed by a freeze that could potentially kill the developing buds or flowers. These concerns call for more research on fruit and nut trees' adaptations to climate change and also on the potential development of new or existing cultivars more resilient to these stressors. Other manifestations of climate change are likely to affect fruit and nut production including changes in rainfall, changes to summer and spring heat or increases in pest pressure due to faster reproduction of ectothermic pest organisms. Increasing temperatures are likely to change pest dynamics in California. For example, generation numbers of codling moth, navel orangeworm, web-spinning spider mites and

European red mites are likely to increase, generating new challenges for pest control. Additional research into agroecological food webs is needed in order to project the implications of climate change for pest pressure with greater certainty (Luedeling et al., 2011b).

Future research must continue to address issues of production, regional adaptability, profitability, cultural and consumer acceptability (Lovell et al., 2018). Economic viability means profitability for farmers and prosperity for rural communities. Cultural acceptability entails meeting the aesthetic, ethical and practical needs of rural communities. Access to improved cultivars, solid production guidelines, reliable product supply, growing consumer demand and sound financial decision support tools are assisting the growth of the speciality crop industry and having positive impacts up and down the supply chain (Mori et al., 2017). Increasing production of tree nuts within alley cropping will help to produce the carbohydrates, proteins and oils that are the basic components of food systems and market supply chains (Kahn et al., 2011; USDA-FS, 2017). Major research frontiers for alley cropping with nut trees include within-system tree diversity, tree crops for food, perennial alley crops and trees for crop facilitation (Wolz and DeLucia, 2018). Future work evaluating the long-term yields and biogeochemical consequences of alley cropping with nut trees are critical to widespread adoption of this transformative agricultural alternative (Wolz et al., 2018).

## 12 Where to look for further information

A number of organizations provide additional resources pertaining to agroforestry, alley cropping and nut trees. Rich sources of information include the USDA National Agroforestry Center, the MU Center for Agroforestry and the Savanna Institute (Table 2).

**Table 2** Alley cropping: additional resources

| Organization/publication | Location of Information |
| --- | --- |
| USDA National Agroforestry Center | https://www.fs.usda.gov/nac/practices/alleycropping.shtml |
| The Center for Agroforestry at the University of Missouri | http://www.centerforagroforestry.org  Alley Cropping: http://www.centerforagroforestry.org/practices/ac.php |
| Savanna Institute | http://www.savannainstitute.org/resources.html |
| Alley Cropping video | http://www.youtube.com/watch?v=b8Kwb5yInPM |
| United Kingdom and Europe | http://www.agforward.eu/index.php/en/silvoarable-agroforestry-in-the-uk.html |
| USDA Natural Resources Conservation Service | www.nrcs.usda.gov/wps/portal/nrcs/detailfull/national/landuse/forestry/sustain/guidance |
| Association for Temperate Agroforestry (AFTA) | http://www.aftaweb.org/about/what-is-agroforestry/alley-croping.html |

# 13 References

Aguilar, F. X., Cernusca, M. M. and Gold, M. A. 2009. Exploratory assessment of consumer preferences for chestnut attributes in Missouri. *HortTechnology* 19(1), 216–23. doi:10.21273/HORTSCI.19.1.216.

Aguilar, F. X., Cernusca, M. M., Gold, M. A. and Barbieri, C. E. 2010. Frequency of consumption, familiarity and preferences for chestnuts in Missouri. *Agroforestry Systems* 79(1), 19–29. doi:10.1007/s10457-009-9266-2.

Anagnostakis, S. L. 2012. Chestnut breeding in the United States for disease and insect resistance. *Plant Disease* 96(10), 1392–403. doi:10.1094/PDIS-04-12-0350-FE.

Bainard, L. D., Koch, A. M., Gordon, A. M. and Klironomos, J. N. 2013. Growth response of crops to soil microbial communities from conventional monocropping and tree-based intercropping systems. *Plant and Soil* 363(1–2), 345–56. doi:10.1007/s11104-012-1321-5.

Bugg, R. L., Sarrantonio, M., Dutcher, J. D. and Phatak, S. C. 1991. Understory cover crops in pecan orchards: possible management systems. *American Journal of Alternative Agriculture* 6(2), 50–62. doi:10.1017/S0889189300003854.

Cardinael, R., Mao, Z., Prieto, I., Stokes, A., Dupraz, C., Kim, J. H. and Jourdan, C. 2015. Competition with winter crops induces deeper rooting of walnut trees in a Mediterranean alley cropping agroforestry system. *Plant and Soil* 391(1–2), 219–35. doi:10.1007/s11104-015-2422-8.

Cartwright, L., Goodrich, N., Cai, Z. and Gold, M. 2017. Using NRCS technical and financial assistance for agroforestry and woody crop establishment through the environmental quality incentives program (EQIP). Agroforestry in Action AF 1016, 4pp. Available at: http://www.centerforagroforestry.org/pubs/NRCS_AgroforestryandWoodyCrop.pdf.

Coggeshall, M. V. 2011a. Use of microsatellite markers to develop new eastern black walnut cultivars in Missouri, USA. *28th International Horticultural Congress*, Lisboa, Portugal. *Acta Horticulturae* 918(1), 221–6. doi:10.17660/ActaHortic.2011.918.27.

Coggeshall, M. V. 2011b. Black walnut: a nut crop for the Midwestern United States. *HortScience* 46(3), 340–2. doi:10.21273/HORTSCI.46.3.340.

Eichhorn, M. P., Paris, P., Herzog, F., Incoll, L. D., Liagre, F., Mantzanas, K., Mayus, M., Moreno, G., Papanastasis, V. P., Pilbeam, D. J., et al. 2006. Silvoarable systems in Europe – past, present and future prospects. *Agroforestry Systems* 67(1), 29–50. doi:10.1007/s10457-005-1111-7.

FAO. 2017. Food and Agriculture Organization of the United Nations, Statistics Division. FAOSTAT. Last Updated: February 2017.

Finney, D. M. and Kaye, J. P. 2017. Functional diversity in cover crop polycultures increases multifunctionality of an agricultural system. *Journal of Applied Ecology* 54(2), 509–17. doi:10.1111/1365-2664.12765.

Flaim, J. 2005. Shepherd farms. In: *The New American Farmer, Profiles of Agricultural Innovation* (2nd edn.). Sustainable Agriculture Network, Beltsville, MD. Available at: https://maawg.files.wordpress.com/2012/07/shepherd-case-study-1.pdf.

Fletcher, E. H., Thetford, M., Sharma, J. and Jose, S. 2012. Effect of root competition and shade on survival and growth of nine woody plant taxa within a pecan (*Carya illinoinensis* K. Koch.) alley cropping system. *Agroforestry Systems* 86(1), 49–60. doi:10.1007/s10457-012-9507-7.

Garrett, H. E., McGraw, R. L. and Walter, W. D. 2009. Alley cropping practices. In: Garrett, H. E. (Ed.), *North American Agroforestry: an Integrated Science and Practice* (2nd edn.). American Society of Agronomy, Madison, WI, pp. 133–62.

Geyer, W. A. and Fick, W. H. 2014. Yield and forage quality of smooth brome in a black walnut alley-cropping practice. *Agroforestry Systems* 89, 107–12.

Gillespie, A. R., Jose, S., Mengel, D. B., Hoover, W. L., Pope, P. E., Seifert, J. R., Biehle, D. J., Stall, T. and Benjamin, T. J. 2000. Defining competition vectors in a temperate alley cropping system in the Midwestern USA: 1. Production physiology. *Agroforestry Systems* 48(1), 25–40. doi:10.1023/A:1006285205553.

Gold, M. A. and Garrett, H.E. 2009. Agroforestry nomenclature, concepts, and practices. In: Garrett, H. E. (Ed.), *North American Agroforestry: an Integrated Science and Practice* (2nd edn). Agronomy Society of America, Madison, WI, pp. 45–56.

Gold, M. A. and Hanover, J. W. 1987. Agroforestry systems for the temperate zone. *Agroforestry Systems* 5(2), 109–21. doi:10.1007/BF00047516.

Gold, M.A., Cernusca, M.M. and Godsey, L. D. 2006. Competitive market analysis: chestnut producers. *HortTechnology* 16(2), 360–9. doi:10.21273/HORTTECH.16.2.0360.

Grant, J., Anderson, K. K., Prichard, T., Hasey, J., Bugg, R. L., Thomas, F. and Johnson, T. 2006. Cover crops for walnut orchards. University of California Agriculture and Natural Resources Publication 21627. Available at: https://anrcatalog.ucanr.edu/pdf/21627e.pdf.

Grauke, L. J. and Thompson, T. E. 2003. Rootstock development in temperate nut crops. In: Janick, J. (Ed.), *Proceedings of the XXVI IHC – Genetics and Breeding of Tree Fruits and Nuts. Acta Horticulturae* 622(622), pp. 553–66. doi:10.17660/ActaHortic.2003.622.59.

Hunt, K. L., Gold, M. A. and Warmund, M. R. 2005. Chinese chestnut cultivar performance in Missouri. In: Abreu, C. G., Rosa, E. and Monteiro, A. A. (Eds), *Proceedings of the Third International Chestnut Congress. Acta Horticulturae* 693(693), pp. 145–8. doi:10.17660/ActaHortic.2005.693.15.

Hunt, K., Gold, M., Reid, W. and Warmund, M. 2012. Growing Chinese chestnut in Missouri. UMCA Agroforestry in Action Guide AF1007-2012, 16pp. Available at: http://extension.missouri.edu/explorepdf/agguides/agroforestry/af1007.pdf.

Inurreta-Aguirre, H. D., Lauri, P. E., Dupraz, C. and Gosme, M. 2018a. Yield components and phenology of durum wheat in a Mediterranean alley-cropping system. *Agroforestry Systems* 92(4), 961–74. doi:10.1007/s10457-018-0201-2.

Inurreta-Aguirre, H. D., Lauri, P. E., Dupraz, C. and Gosme, M. 2018b. Impact of tree root pruning on yield of durum wheat and barley in a Mediterranean alley-cropping system. *Proceedings Fourth European Agroforestry Conference*, Nijmejen, Therapeutic Netherlands. Available at: https://www.researchgate.net/publication/325597457.

Johnson, R. 2014. Fruits, vegetables, and other specialty crops: selected farm bill and other federal programs. Rep. No. R42771. Congressional Research Service, Washington DC, 51pp. Available at: http://www.nationalaglawcenter.org/wp-content/uploads/assets/crs/R42771.pdf.

Jose, S. 2009. Agroforestry for ecosystem services and environmental benefits: an overview. *Agroforestry Systems* 76(1), 1–10. doi:10.1007/s10457-009-9229-7.

Jose, S., Gillespie, A. R. and Pallardy, S. G. 2004. Interspecific interactions in temperate agroforestry. *Agroforestry Systems* 61(1), 237–55.

Jose, S., Holzmueller, E. J. and Gillespie, A. R. 2009. Tree-Crop Interactions in temperate agroforestry. In: Garrett, H. E. (Ed.), *North American Agroforestry: an Integrated Science and Practice* (2nd edn.). American Society of Agronomy, Madison, WI, pp. 57–74.

Jose, S., Gold, M. A. and Garrett, H. E. 2018. Temperate agroforestry in the United States: current trends and future directions. In: Gordon, A. M., Newman, S. M. and Coleman, B. (Eds), *Temperate Agroforestry Systems* (2nd edn.). CAB International, Boston, MA, pp. 51-71. Chapter 3.

Kahn, P. C., Molnar, T., Zhang, G. C. and Funk, C. R. 2011. Investing in perennial crops to sustainably feed the world. *Issues in Science and Technology* 27(4), 75-81.

Krebs, J. and Bach, S. 2018. Permaculture–scientific evidence of principles for the agroecological design of farming systems. *Sustainability* 10(9), 3218-42. doi:10.3390/su10093218.

Kremer, R. J. and Kussman, R. D. 2011. Soil quality in a pecan-kura clover alley cropping system in the Midwestern USA. *Agroforestry Systems* 83(2), 213-23. doi:10.1007/s10457-011-9370-y.

Lacombe, S., Bradley, R. L., Hamel, C. and Beaulieu, C. 2009. Do tree-based intercropping systems increase the diversity and stability of soil microbial communities? *Agriculture, Ecosystems and Environment* 131(1), 25-31.

Lelle, M. A. and Gold, M. A. 1994. Agroforestry systems for temperate climates: lessons from Roman Italy. *Forest and Conservation History* 38(3), 118-26. doi:10.2307/3983919.

Lepofsky, D. 2009. The past, present, and future of traditional resource and environmental management. *Journal of Ethnobiology* 29(2), 161-6. doi:10.2993/0278-0771-29.2.161.

Lovell, S. T., Dupraz, C., Gold, M., Jose, S., Revord, R., Stanek, E. and Wolz, K. J. 2018. Temperate agroforestry research - considering multifunctional woody polycultures and the design of long-term field trials. *Agroforestry Systems* 92(5), 1397-415. doi:10.1007/s10457-017-0087-4.

Low, S. A. and Vogel, S. J. 2011. Direct and intermediated marketing of local foods in the United States. ERR-128. United States Department of Agriculture, Economic Research Service. Available at: https://www.ers.usda.gov/webdocs/publications/44924/8276_err128_2_.pdf?v=41056.

Luedeling, E., Girvetz, E. H., Semenov, M. A. and Brown, P. H. 2011a. Climate change affects winter chill for temperate fruit and nut trees. *PLoS ONE* 6(5), e20155. doi:10.1371/journal.pone.0020155.

Luedeling, E., Steinmann, K. P., Zhang, M., Brown, P. H., Grant, J. and Girvetz, E. H. 2011b. Climate change effects on walnut pests in California. *Global Change Biology* 17(1), 228-38. doi:10.1111/j.1365-2486.2010.02227.x.

Malézieux, E. 2012. Designing cropping systems from nature. *Agronomy for Sustainable Development* 32(1), 15-29. doi:10.1007/s13593-011-0027-z.

Martin, A. R. and Isaac, M. E. 2018. Functional traits in agroecology: advancing description and prediction in agroecosystems. *Journal of Applied Ecology* 55(1), 5-11. doi:10.1111/1365-2664.13039.

Miller, M., Williams, B., Raboin, M., Carusi, C., McNair, R. and Bauer, L. 2016. Growing Midwestern tree nut businesses: five case studies. CIAS, Madison, WI. Available at: https://www.cias.wisc.edu/wp-content/uploads/2016/11/ciasgrowingmidwesttreenutfinallowres.pdf.

Molnar, T. J. and Capik, J. M. 2012. Advances in hazelnut research in North America. *Acta Horticulturae* 940(940), 57-65. doi:10.17660/ActaHortic.2012.940.6.

Molnar, T. J., Goffread, J. C. and Funk, C. R. 2005. Developing hazelnuts for the Eastern United States. In: Tous, J., Rovira, M. and Romero, A. (Eds), *Proceedings of the VIth*

*International Conferences on Hazelnut. Acta Horticulturae* 686, 609–18. Illinois State Historical Society.

Molnar, T. J., Khan, P. C., Ford, T. M., Funk, C. J. and Funk, C. R. 2013. Tree crops, a permanent agriculture: concepts for the past for a sustainable future. *Resources* 2(4), 457–88. doi:10.3390/resources2040457.

Moreno, G., Aviron, S., Berg, S., Crous-Duran, J., Franca, A., García de Jalón, S., Hartel, T., Mirck, J., Pantera, A., Palma, J. H. N., et al. 2018. Agroforestry systems of high nature and cultural value in Europe: provision of commercial goods and other ecosystem services. *Agroforestry Systems* 92(4), 877–91. doi:10.1007/s10457-017-0126-1.

Mori, G. O., Gold, M. A. and Jose, S. 2017. Specialty crops in temperate agroforestry systems: sustainable management, marketing and promotion for the Midwest region of the U.S.A. In: Montagnini, F. (Ed.), *Integrating Landscapes: Agroforestry for Biodiversity Conservation and Food Sovereignty. Advances in Agroforestry* 12, 331–66. doi:10.1007/978-3-319-69371-2_14.

Muschler, R. G. 2015. Agroforestry: essential for sustainable and climate-smart land use? In: Pancel, L. and Kohl, M. (Eds), *Tropical Forestry Handbook* (2nd edn.). Springer-Verlag, Berlin Heidelberg, pp. 2013–116.

Nair, P. K. R. 1993. *An Introduction to Agroforestry*. Kluwer Academic Publishers, Dordrecht, Netherlands, 505pp.

Nicholls, C. I., Altieri, M. A. and Vazquez, L. 2016. Agroecology: principles for the conversion and redesign of farming systems. *Journal Ecosys Ecography* S5, 010. doi:10.4172/2157-7625.S5-010.

Pantera, A., Burgess, P. J., Mosquera Losada, R., Moreno, G., Lopez-Dıaz, M. L., Corroyer, N., McAdam, J., Rosati, A., Papadopoulos, A. M., Graves, A., et al. 2018. Agroforestry for high value tree systems in Europe. *Agroforestry Systems* 92(4), 945–59. doi:10.1007/s10457-017-0181-7.

Reid, W., Coggeshall, M. V. and Hunt, K. L. 2004. Cultivar evaluation and development for black walnut orchards. In: Michler, C. H., Pijut, P. M., Van Sambeek, J. W., Coggeshall, M. V., Seifert, J., Woeste, K., Overton, R. and Ponder Jr., F. (Eds), *Black Walnut in a New Century. Proceedings of the 6th Walnut Council Research Symposium*, 25–28 July 2004, Lafayette, IN. Gen. Tech. Rep. NC-243. United States Department of Agriculture, Forest Service, North Central Research Station, St. Paul, MN, 188pp.

Reid, W., Coggeshall, M. V., Garrett, H. E. and Van Sambeek, J. W. 2009. Growing black walnut for nut production. 'Agroforestry in Action' publication AF1011-2009. University of Missouri Center for Agroforestry, 16pp. Available at: http://extensio n.missouri.edu/explorepdf/agguides/agroforestry/af1011.pdf.

Reijntjes, C., Haverkort, B. and Waters-Bayer, A. 1992. *Farming for the Future: an Introduction to Low-External-Input and Sustainable Agriculture*. Macmillan, London, UK.

Rigueiro-Ródriguez, A., Fernandez-Nunez, E., Gonzalez-Hernandez, P., McAdam, J. H. and Mosquera-Losada, M. R. 2009. Agroforestry systems in Europe: productive, ecological and social perspectives. In: Rigueiro-Rodríguez, A., McAdam, J. and Mosquera-Losada, M. R. (Eds), *Agroforestry in Europe: Current Status and Future Prospects, vol. 6*. Springer, Dordrecht, pp. 43–65.

Rossier, C. and Lake, F. 2014. Indigenous traditional ecological knowledge in agroforestry. Agroforestry Notes 44. United States Department of Agriculture, National Agroforestry Center, Lincoln, NE, 8pp.

Scott, R. and Sullivan, W. C. 2007. A review of suitable companion crops for black walnut. *Agroforestry Systems* 71, 185–93. doi:10.1007/s10457-007-9071-8.

Shapiro-Ilan, D. I., Gardner, W. A., Wells, L. and Wood, B. W. 2012. Cumulative impact of a clover cover crop on the persistence and efficacy of *Beauveria bassiana* in suppressing the pecan weevil (Coleoptera: Curculionidae). *Environmental Entomology* 41(2), 298–307. doi:10.1603/EN11229.

Shelton, A. C. and Tracy, W. F. 2017. Cultivar development in the U.S. public sector. *Crop Science* 57(4), 1823–35. doi:10.2135/cropsci2016.11.0961.

Smith, J. R. 1950. *Tree Crops: A Permanent Agriculture*. Devin-Adair, New York, NY.

Stamps, W. T., McGraw, R. L., Godsey, L. and Woods, T. L. 2009. The ecology and economics of insect pest management in nut tree alley cropping systems in the Midwestern United States. *Agriculture, Ecosystems and Environment* 131(1–2), 4–8. doi:10.1016/j.agee.2008.06.012.

Thevathasan, N. V., Gordon, A. M., Simpson, J. A., Reynolds, P. E., Price, G. and Zhang, P. 2004. Biophysical and ecological interactions in a temperate tree-based intercropping system. *Journal of Crop Improvement* 12(1–2), 339–63.

Udawatta, R. P. and Jose, S. 2012. Agroforestry strategies to sequester carbon in temperate North America. *Agroforestry Systems* 86(2), 225–42. doi:10.1007/s10457-012-9561-1.

USDA-ERS. 2015. Trends in U.S. local and regional food systems: report to Congress. Administration Publique No. 068. United States Department of Agriculture, Economic Research Service, Washington DC, 47pp.

USDA-ERS. 2018. National count of farmers market directory listings: Graph 1994–2017. Available at: https://www.ams.usda.gov/sites/default/files/media/NationalCountoffFMDirectory17.JPG.

USDA-FS. 2017. Agroforestry: enhancing resiliency in U.S. agricultural landscapes under changing conditions. Schoeneberger, M. M., Bentrup, G. and Patel-Weynand, T. (Eds). Gen. Tech. Report WO-96. United States Department of Agriculture, Forest Service, Washington DC. doi:10.2737/WO-GTR-96.

Valdivia, C., Barbieri, C. and Gold, M. A. 2012. Between forestry and farming: policy and environmental implications of the barriers to agroforestry adoption. *Canadian Journal of Agricultural Economics/Revue Canadienne d'Agroeconomie* 60(2), 155–75. doi:10.1111/j.1744-7976.2012.01248.x.

Vandermeer, J. 1995. The ecological basis of alternative agriculture. *Annual Review of Ecology and Systematics* 26(1), 201–24. doi:10.1146/annurev.es.26.110195.001221.

Van Sambeek, J. and Reid, W. 2017. A double row alley-cropping system for establishing nut orchards. *MNGA (Missouri Nutrition Growers Association) Newsletter* 17(4), 11–4.

Walter, W. D., Jose, S. and Zamora, D. 2015. Alley cropping. In: Gold, M. A., Cernusca, M. M. and Hall, M. (Eds), *Training Manual for Applied Agroforestry Practices – 2015 Edition*. MU Center for Agroforestry, pp. 31–49. Available at: http://www.centerforagroforestry.org/pubs/training/index.php.

Wechsler, D. 2005. Hopeton farms. In: *The New American Farmer, Profiles of Agricultural Innovation* (2nd edn.). Sustainable Agriculture Network, Beltsville, MD. Available at: https://www.sare.org/Learning-Center/Books/The-New-American-Farmer-2nd-Edition.

Wells, L., Conner, P., Brock, J. and Hudson, W. 2018. Organic pecan production. UGA Coop Extension Bulletin 1493. Available at: https://secure.caes.uga.edu/extension/publications/files/pdf/B%201493_1.PDF.

Wilson, M. H., Lovell, S. T. and Carter, T. 2018. *Perennial Pathways: Planting Tree Crops. Designing and Installing Farm-Scale Edible Agroforestry*. Savanna Institute, Madison, WI.

Wolz, K. J. and DeLucia, E. H. 2018. Alley cropping: global patterns of species composition and function. *Agriculture, Ecosystems and Environment* 252, 61–8. doi:10.1016/j.agee.2017.10.005.

Wolz, K. J., Lovell, S. T., Branham, B. E., Eddy, W. C., Keeley, K., Revord, R. S., Wander, M. W., Yang, W. H. and DeLucia, E. H. 2017. Frontiers in alley cropping: transformative solutions for temperate agriculture. *Global Change Biology* 24, 883–94.

Wolz, K. J., Branham, B. E. and DeLucia, E. H. 2018. Reduced nitrogen losses after conversion of row crop agriculture to alley cropping with mixed fruit and nut trees. *Agriculture, Ecosystems and Environment* 258, 172–81. doi:10.1016/j.agee.2018.02.024.

Wood, B. W., Grauke, L. J. and Bock, C. H. 2016. The pecan provenance collection at Byron Ga – a unique resource for the long-term survival of the US industry. *Pecan Grower* 27, 28–46.

Zhaohua, Z., Maoyi, F. and Stastry, C. B. 1991. Agroforestry in China – an overview. In: Zhaohua, Z., Mantang, C., Shiji, W. and Youxu, J. (Eds), *Agroforestry Systems in China*. Chinese Academy of Forestry and International Development Research Center, Canada.

www.ingramcontent.com/pod-product-compliance
Lightning Source LLC
Chambersburg PA
CBHW050533270326
41926CB00015B/3208